Infinite Universe

Maciej B. Szymanski

Simplicity is the ultimate sophistication.
—Leonardo da Vinci

Canadian Cataloguing in Publication Data
Library and Archives Canada Cataloguing in Publication

Szymanski, Maciej B. (Maciej Boleslaw), author
Infinite universe / Maciej B. Szymanski.

Includes bibliographical references, glossary and index.
ISBN 978-0-9951680-0-8 (Canada)

1. Cosmology. 2. Universe. 3. Physics.
4. Gravitation. 5. Speed of Light. I. Title.

QP981.S99 2016   523.1   C2016-902409-1

Library of Congress Control Number: 2016909490
ISBN 978-1-5308-7714-0 (USA)

First Edition

Copyright © 2016 by Maciej B. Szymanski.
All rights reserved.
No part of this book may be used or reproduced in any manner whatsoever without written permission except in the case of brief quotations embodied in critical articles and reviews.

For permission and other information email
maciej.b.szymanski@gmail.com.

Published by Maciej B. Szymanski (mbs)
www.maciejbszymanski.com

CreateSpace Independent Publishing Platform
North Charleston, SC
Printed by CreateSpace, an Amazon company

To my grandsons
Aiden and Tristen

CONTENTS

ABBREVIATIONS vi

PREFACE vii

A NOTE TO THE NONSPECIALIST READER xii

CHAPTER I: INFINITE TIME 1

1 Fundamental Premises of Standard Cosmological Model 1
2 Infinite-Time Hypothesis 8
3 Cosmological Consequences of Infinite Time 11

CHAPTER II: DEHUMANIZATION OF NATURE 15

4 Background 15
5 Some Crucial Issues in the Philosophy of Physics 16
6 Fundamental-Law Hypothesis 17
7 Fundamental Laws of Physics 22
8 Correct Statement of a Fundamental Law of Physics 26
9 Action-at-a-Distance 28
10 Simplicity-of-Execution Principle 33
11 Second Law of Thermodynamics 37
12 Other Laws in Physics 40
13 Space and Time 43
14 Physics and Mathematics 47
15 Physics in Trouble 54

CHAPTER III: GRAVITATION 56

16 Background 56
17 Finite Range of Gravity 57
18 Speed of Gravity 63
19 Gravity (Gravitational Field) 68
20 Gravitational Interactions 72
21 Gravity-Conservation Law 76
22 Gravity-Force Equation 79

23	Gravity versus Electrostatic Force	89
24	The Influence of Distant Masses	91
25	The Force of Antigravity	96
26	Equivalence Principle	98
27	Range of Gravitational Interactions	103

CHAPTER IV: PHOTONS 107

28	Background	107
29	Photon Energy	108
30	Maintaining the Speed of the Photon	109
31	Steady-State Speed of Radiation	112
32	The Speed of the Photon	118
33	The Finite Speed of the Photon	124
34	Maxwell's Aether and Einstein's Vacuum	125

CHAPTER V: UNIVERSE 126

35	Summary of Conclusions Drawn in Chapters I through IV	126
36	Background	128
37	The Insufficient-Mass Conclusion	129
38	The Structure of the Universe	132
39	Cosmic Void and Star Horizons	135
40	Mass Density in the Universe	137
41	The Life of a Star	139
42	Perpetuum Mobile	142
43	Distribution of Galaxies in Galaxy Filament	143
44	Cosmological Redshifts	144
45	Cosmic Microwave Background	151
46	Summary of the GPU Cosmology	154

APPENDIX A: Glossary of Terms and Concepts
APPENDIX B: Index
APPENDIX C: Bibliography

ABBREVIATIONS

CMB	- cosmic microwave background
FLH	- fundamental-law hypothesis
FLP	- fundamental law of physics (nature)
GF	- galaxy filament
GPU	- Gravitation Photons Universe
RCoPM	- reduced components of photon momenta
RVU	- representative volume of the universe
SLT	- second law of thermodynamics
VO&VC	- very old and very cold (stars/galaxies)

PREFACE

The purpose of this book is to present and discuss the cosmology of the universe that is infinite in time and to show such a universe to be a rational alternative to the widely accepted STANDARD COSMOLOGICAL (BIG-BANG) MODEL[a]. Infinite time implies that the universe is not expanding. I will discuss this implication in section 3. As the expansion of the universe constitutes the basis of the standard cosmological model, one might reason that a discussion of the cosmology of a nonexpanding universe is superfluous. That, however, would be a biased stance since, as I will point out in section 1, there is no direct evidence of any kind that would support the notion of an expanding universe. The mere lack of such evidence cannot, of course, be used to argue that the standard model is erroneous. Nevertheless, that lack of direct evidence presents a compelling argument for the need to consider alternate cosmologies.

The cosmology of the infinite-in-time universe will be construed from three hypotheses, numbered (i), (ii), and (iii). First, it will be postulated that the perfect cosmological principle holds. That principle states that the universe is homogeneous and isotropic on a large space scale and on a large time scale, which implies that the universe is infinite in time and nonevolving.[b] Henceforth, I will refer to the premise of infinite time, which is a direct consequence of the perfect cosmological principle, as (i) the infinite-time hypothesis, keeping in mind that the homogeneity and isotropy of the universe on a large space scale are also assumed. As I will discuss in section 3, the infinite-time hypothesis implies that the universe is also spatially infinite and that the range of gravitational interaction (the "range of gravity") is finite.

---

[a] Most of the important terms and concepts that a nonspecialist in physics or cosmology may be unfamiliar with are typed in small caps where they first appear and are briefly explained in appendix A. Some of such terms and concepts are explained directly in the text.

[b] The perfect cosmological principle can be more accurately stated as: the appearance of the universe does not change in space and over time. That statement explicitly excludes the possibility of an expansion of the universe. In the standard model, the appearance of the universe has been continuously changing since the big-bang "explosion," for instance, the average mass density in the universe has been decreasing due to the expansion of space.

Second, the foundations of classical (i.e., non-QUANTUM) PHYSICS will be laid down and the fundamental laws of nature (physics) identified. Those laws will be identified based on the following hypothesis: (ii) fundamental laws of nature do exist independent of the human mind. They will suggest physics at the fundamental level to be simple in the extreme.

The third hypothesis concerns a finite range of gravity, which is implied by the infinite-time hypothesis. It will be introduced in the form of the following postulate: (iii) the range of gravitational interactions of an object increases with the contents of its mass. Herein, "mass" means "intrinsic mass." The meaning of intrinsic mass is explained in the last paragraph of section 16.

In developing the cosmological background to the conception of an infinite-in-time universe, I will suggest, based on the second and third hypotheses, that the speed of light has to be constant and independent of the speed of the light-emitting source only when subject to gravitation. That, in turn, will provide a background to the explanation of the physics underlying the famous limit on the speed of the motion of matter and energy: "nothing can travel faster than light." However, attempting to apply similar explanation to the speed of information transfer, where the information is transferred by an interaction only—that is, transferred without matter or energy involved—will fail to identify any speed limit. This means that the interactions that involve neither matter nor energy can travel faster than light.

I will also demonstrate that a simple and self-consistent law of gravitation can be derived based on hypotheses two and three. After some averaging of fundamental variables, that law will turn out to be the same as Newton's law of gravitation, except for a first-approximation modification introduced to account for a finite range of gravity. The proposed modification is quantitatively negligible when the results of the observations of the sun-planet gravitational interactions are considered or when the results of gravitational experiments carried out in laboratories are examined. (Note that Newton derived his law of gravitation from such observations and experiments.) Furthermore, an explanation of the widely pondered conundrum of the ratio of the magnitude of electrostatic force to gravity force will be proposed based on hypotheses two and three. It will involve the consideration of a force of antigravity. I believe that explanation to be crucial to the understanding of gravitation.

Finally, a conceptual cosmological model of the universe will be constructed from hypotheses one, two, and three. It will be consistent with the fundamentals of classical physics and in agreement with the results of key astronomic observations. The term "conceptual" is meant to emphasize that it is not my intent to advance that model to a level that would allow for making quantitative predictions of the dynamics of the universe. Making quantitative predictions would require the construction of a mathematical framework over the physical model of the universe. Rather, my objective here is to explain how the infinite-in-time universe would be expected to work in nonquantitative terms.

While hypothesis two, which I will discuss in detail in chapter II, seems to be in disagreement with the currently prevailing understanding of the fundamentals of classical physics, I will not in any way suggest that modern physics has been developed using an incorrect or inadequate approach. On the contrary, I will argue that applied physics has been developed using the most suitable and highly effective approach, which comprises the development of mathematical models designed to describe and predict physical phenomena. However, within the framework of modern physics, it appears to be very difficult to develop a verifiable cosmological model that would explain astronomic observations in physical, as opposed to mathematical, terms. I will show that accepting an alternate understanding of the fundamentals of classical physics allows for the development of a self-consistent cosmological model that appears to be as credible as the standard cosmological model and fully verifiable, at least in principle. "Verifiable in principle" means that given adequate technical means, the proposed model could be verified today without the need to venture into the inaccessible past or the future, which would be required to verify the standard model.

It also appears very difficult, if not impossible, to resolve some problems associated with the standard cosmological model, such as the undetectability of DARK MATTER and DARK ENERGY or the HORIZON and FLATNESS PROBLEMS. Alan Guth's hypothesis of the EARLY INFLATION OF THE UNIVERSE[1] was put forward with the objective of removing the horizon and flatness problems. However, that (unverifiable) hypothesis has been questioned by Paul Steinhardt,[2] Roger Penrose,[3] and other physicists, so those two problems cannot be said to have been resolved. The cosmological model put forward in this book is free of those kinds of problems.

The cosmology of the infinite-in-time universe is referred to as the "GPU cosmology." The acronym GPU stands for "Gravitation Photons Universe," which are the key elements of the proposed cosmological model. Hypotheses one through three, together with the well-established premises of classical physics such as the second law of thermodynamics or the conservation laws, form the GPU theory that underlies the GPU cosmological model.[c] The way the infinite universe works is very simple, according to that model. It can be explained by a physics teacher at the senior-year-of-high-school level. Perhaps the most salient feature of the GPU theory is the complete dehumanization of nature (physics).

With regard to the actual astronomic observations, the GPU model goes beyond the standard model. While the GPU model explains the results of astronomic observations that are often explained differently by the standard model—for example, the OLBERS'S PARADOX, COSMIC MICROWAVE BACKGROUND, or COSMOLOGICAL REDSHIFTS—it also explains observations that cannot be explained by the standard model without introducing unverifiable AD HOC HYPOTHESES. Examples of such observations include the clustering of visible GALAXIES, the types and ages of VOID GALAXIES, the existence of COSMIC VOIDS, and the apparent weakness of the force of gravity.

Most of popular-physics books describe some well-established and generally accepted central aspects of physics in a simplified, typically nonmathematical way. In this book, when I introduce a new aspect of physics, I will attempt to explain it in full without any simplifications and without the use of advanced mathematics. Neither simplifications nor advanced math will be needed. (As I will point out in chapter II, my view is that there is no room for mathematics at *the fundamental level of physics*. This view is shared by many philosophers of science and a number of physicists, including some Nobel Prize winners.) There are a few high-school level equations in this book, but their understanding is

---

[c] By a "theory," I mean here a set of scientific hypotheses/principles that can underlie more than one component of science. By a "scientific model," I mean the application of a scientific theory to the description of a particular component of science.

not critical to the understanding of the GPU cosmology since the implications of those equations can also be explained in simple words. As a result, I think of this book not as a popular-science book but rather as a book on the following concepts concerning physics and cosmology: the fundamental laws of physics, time and space, gravitation, the speed of light, and the universe.

There is also another reason that this book cannot be regarded as a popular-science book: its contents are based on some new and reinvented (i.e., considered in the past but currently abandoned) physics concepts, so it cannot be thought of as a book that presents well-established aspects of physics. The reinvented and new physics concepts discussed in this book include a universe that is infinite in time and space, the applicability of the second law of thermodynamics to the distribution of mass, a finite range of gravity, a nonmathematical gravitational field, and the concept of a true vacuum that refers to those regions of space where no potential for gravitational interactions exists.

The contents of this book were adopted from three papers[4][5][6] published by the author between 2010 and 2012, with some suggestions, conclusions and clarifications added and a few earlier errors corrected.

A NOTE TO THE NONSPECIALIST READER

I have attempted to design this book so that the cosmology of the infinite universe can be understood and appreciated by the curious nonspecialist. That, however, does not mean that the book has become easy to read. I expect that the nonspecialist will have to make some effort to fully understand and appreciate its contents. That is not because the concepts presented in this book are complex, or technically difficult to grasp, or explained using advanced mathematics. Rather, the nonspecialist will have to make an effort to fully understand the contents of this book because of a large number of scientific terms and concepts referred to in the discussions that s/he may be unfamiliar with (e.g., the EQUIVALENCE PRINCIPLE, the CAUSALITY PRINCIPLE, THERMAL RADIATION, cosmic voids, cosmic microwave background, or cosmological redshifts). To aid that effort, I have provided brief explanations of many scientific terms and concepts referred to throughout the book in appendix A. Still, the reader may find Wikipedia articles or the websites of NASA and some universities highly useful. Those articles and websites often provide simple yet more elaborate explanations of the scientific terms and concepts introduced and discussed in this book.

As I pointed out in the preface, the chief objective of this book is to describe the cosmology of the infinite-in-time universe. That, however, will only be done in chapter V, in which the key aspects of the GPU cosmology will be elaborated on. There is a lot of information before chapter V, discussed in chapters I through IV, in which the physical concepts and principles underlying the GPU theory are addressed. To assist the reader who is interested in the model of the infinite-in-time universe only, I state at the beginning of chapters II, III, and IV which sections have to be read in order to appreciate the key features of the infinite-in-time universe. The reader who would like to know about the simplicity and thus the immense beauty of nature and the universe should read all of the sections.

Still, you may ask why it takes so long to get to the description of the infinite universe that is presented in the last chapter only. The answer is that in chapters I through IV, the questions central to any cosmological model are addressed to show that the GPU model has robust theoretical support. Those chapters form the background of the GPU cosmology.

Section 3 in chapter I contains a limited overview of the cosmology of the infinite universe. The fundamental laws of physics that will allow for explaining how the infinite universe works are identified in chapter II. Gravitation and ELECTROMAGNETIC RADIATION (light) are the central physical phenomena, upon which the GPU cosmology is founded. Those are discussed in chapters III and IV, respectively.

As I proposed above, you may decide to read only the sections that are directly related to the GPU model. If you decide to read all sections, which I recommend, you may become frustrated by your inability to quickly relate some of the presented discussions to the cosmology of the infinite universe, the appreciation of which is your ultimate goal. If you do get frustrated, it will be because of my failure to present you with the rationale for introducing the various aspects of physics discussed in this book. To reduce the chance of me failing and you getting frustrated, I provide in the following paragraphs some clarifications as to why the presented discussions are not only relevant but in fact necessary to substantiate the viability of the proposed cosmological model.

In section 1, I will justify the need to consider the cosmology of an infinite-in-time and nonexpanding universe which, of course, is in basic disagreement with the standard cosmological model. Without that justification, I would question myself the need to discuss an infinite-in-time universe. In section 2, I will assume that time is infinite, and in section 3, I will conclude some cosmological consequences of that assumption. I will need to conclude those consequences because they will become key elements of the GPU cosmology. The only background required to identify those elements is the infinite-time hypothesis.

The discussions presented in chapter II revolve around the fundamental laws of physics. You may ask how those discussions relate to the cosmology of the infinite universe, which is supposed to be the subject of this book. You may even get angry and say, "I didn't get this book to learn about the fundamental laws of physics! I got it to learn about the infinite universe!" Sorry, but you actually should learn about the fundamental laws of physics. Why? It is because while reading this book, you will most likely ponder the widely known and accepted by most physicists big-bang cosmological model. The big-bang model is based on a solution to the equations of GENERAL RELATIVITY, which I will argue is not a fundamental law of nature. I believe that general relativity, which is based on the concept of SPACE-TIME, exists in the physicist's mind only. On the contrary, the cosmology of the infinite universe discussed in this book is based on the laws of nature that exist independent of the physicist's mind. While I certainly recognize how ingenious and useful the concept of space-time is, I have not seen any proof or even the slightest hint that it reflects a physical reality. Since I would like to construct a cosmological model that does reflect physical reality, I am unable to incorporate the concept of space-time into it. As a result, no biases present in the physicist's mind will be embedded in the model of the infinite universe. You need to appreciate that to decide which cosmological model appeals to you more.

The fundamental laws of physics will be identified in section 7, and the simplicity of their execution will be stated, as a principle, in section 10. That principle will be important to the explanations of gravitational interactions, the principle of equivalence, the apparent weakness of the force of gravity, and the speed of light. Gravitational interactions and the equivalence principle are fundamental to any cosmological model. The fundamental laws will be identified based on the assertion that nature does not do mathematics, measure physical quantities, or make decisions. Nor does nature have a need or desire to describe or predict physical phenomena, which is the leading goal of the modern physicist. I will discuss those assertions in detail in section 6. In chapter II, I will discuss the concepts of space and time, and the use of mathematics in advancing physics. Those discussions are intended to support the identification of the fundamental laws of nature, which will be used to explain how the infinite universe works.

In chapter III, I will discuss gravitational interactions. The discussion is rather extensive. It is necessary to show that the notion of a finite range of gravity is feasible. A finite range of gravity is a key element of the GPU cosmology. I have to examine it from various perspectives because it is in sharp disagreement with the views of nearly all physicists since Newton day. The discussion presented in section 20 is of particular importance, as the conclusion of the negligible strength of indirect gravitational interactions will allow for explanations of the apparent weakness and finite strength of the force of gravity, and the equivalence principle. (Of course, I will carefully explain what "indirect" gravitational interaction means.) What I am saying here is that while the discussion of gravitational interactions may seem to be too long and, at times, difficult to follow, I am presenting it for a good reason. No cosmological model can be considered complete, even at the conceptual stage, if it does not explicitly account for gravitational interactions.

The discussion on the speed of light (chapter IV) is also extensive and rather uncommon in most publications on cosmology. In this regard, it needs to be said that the primary purpose of that discussion is not to explain the background to the most famous scientific assumption, "no matter or energy can travel faster than the speed of light". The primary purpose of presenting chapter IV is to show that the speed of light is constant and independent of the speed of the emitting source *only when the light is subject to gravitation*. New stars can then be born out of the thermal radiation emitted by other stars, which is a prerequisite of the GPU cosmology.

Throughout the book I often discuss electromagnetic radiation (more accurately, "thermal radiation") not only in the context of fuel for the formation of new stars but also in the context of cosmological redshifts—which, in the standard cosmological model, are assumed to prove the expansion of the universe—and in the context of the speed of light. Thermal radiation is a critical component of the GPU cosmology. I want you to feel at ease when reading about thermal radiation. There is nothing difficult to understand about it in the context of this book. Put simply, any object, whether it is a spoon, a moon, or a goon, with a temperature above absolute zero emits thermal radiation. That radiation represents thermal energy converted into electromagnetic energy, which comprises PHOTONS. Sunlight is the most familiar example of thermal radiation. When pondering thermal radiation, keep in mind that emitted photons may contain different amounts of energy, from extremely small amounts to extremely high amounts. The majority of the photons in the universe comprise either less or more energy than visible-light photons, and those photons cannot be seen by the human eye. But they are there, for instance, high-energy X-ray or low-energy radio-wave photons. What is most remarkable about thermal radiation is that the intensity of radiation is independent of the composition of the radiating object. It depends on its surface temperature only. Consider, for instance, a hot stove. The intensity of the thermal radiation will be the same regardless of whether the stove is made out of iron, copper, ceramics, or brick as long as its surface temperature, which is approximately the same as the temperature inside the stove, remains constant. The above comments present a limited and simplified explanation of thermal radiation. There are, of course, more complex aspects of it. For the purpose of this book, however, that explanation is entirely sufficient.

Throughout the book I also often refer to the second law of thermodynamics (the SLT). Again, in the context of this book, there is nothing difficult to understand about that law. It simply states that nature eradicates any nonequilibrium between two (connected) locations containing different amounts of physical entities. For instance, place a hot wood stove in the center of a cold room. When you do, there will be much more heat energy at the center of the room than along the walls—that is, a state of nonequilibrium in the distribution of the heat energy will exist. Nature will then eradicate that nonequilibrium. As a result, the temperatures in the room, including those along the walls, will increase in the process of heat convection, while the temperature of the stove will decrease. The various locations in the room are connected by air, which is the carrier of heat energy. The same thing would happen if you heated up one end of a metal rod. The metal would be the carrier of heat energy. If you stop heating the rod, the temperatures across it will equalize after a period of time—that is, the thermal nonequilibrium will be eradicated in the process of heat conduction. That is how the SLT works, not only with respect to nonequilibrium in the distribution of heat energy but also with respect to any nonequilibrium in the distribution of a physical entity. Think, for instance, about the water pressure in a house's water-supply pipe (about 50 psi) and about what happens if a kitchen faucet is open. The water pressure at the faucet's spout is lower, at an atmospheric pressure of about 15 psi at sea level. When you see water flowing out of a faucet, you are seeing the SLT at work. It eradicates the nonequilibrium in water pressures. Actually, it tries to eradicate the nonequilibrium, which is not possible because of the conflict with the constant pressure in the house's water-supply pipe that is maintained by the water-supply utility. The constant pressures in the water-supply pipe and at the faucet spout prevent the water-pressure nonequilibrium from being eradicated as a constant degree of nonequilibrium is forced. When the degree of nonequilibrium is constantly maintained (forced), the resulting process is called a STEADY-STATE thermodynamic process.

It is useful to think of the SLT in terms of order and disorder. For instance, when you place a hot stove in the center of a room, the heat energy in the room becomes highly ordered—that is, there is a lot of heat energy at the center of the room and much less in the remainder of the room. By the next day, the wood would burn out, the stove would cool down, and the temperature would equalize all across the room. The distribution of heat energy would become highly disordered. The SLT *always* drives a system toward a more disordered state. Ask not why that is. Nobody knows. Ask not why or how any other fundamental law of nature is executed. Most likely, those questions have no answers and thus are irrelevant.

I will introduce the concept of a gravitational field that is entirely immaterial—that is, it comprises neither matter nor energy, except that it contains a *potential* energy. (A "field" in physics is something that exists in space or space-time that you cannot see. It affects the behavior of matter and energy.) Yet the field introduced in this book is as real as any other physical entity. While you cannot see it, you can detect it by inserting into the space a piece of matter or energy and observing its reaction. If there is no reaction, then no gravitational field exists. You may get confused if you are familiar with the common concept of a field in physics, for instance, the electromagnetic field. Those fields comprise numbers. There are no numbers in the gravitational field that I will introduce. It is a new concept, and you will need to take it for what it is.

There are numerous references to scientific papers and books made throughout this book. In some cases, those references are made to indicate that some physicists, cosmologists, or philosophers of science may have opinions different from those expressed herein. In many cases, the references to papers and other books are made to indicate theoretical or experimental support for the ideas and conclusions put forward in this book. Unless you want to check how strong that support is or how strong the opposition to an idea or conclusion proposed herein is, there should be no compelling reason for you to go to the trouble of searching for, obtaining, and then reading a referenced book or paper. In any case, there are two things you should know about the references I chose to quote: First, most of the referenced books and papers can be read and understood by a nonspecialist. Second, many of the referenced books and papers are available for free on the Internet.

Infinite Universe

So far as hypotheses are concerned, let no one expect anything certain from astronomy, which cannot furnish it, lest he accept as the truth ideas conceived for another purpose, and depart from this study a greater fool than when he entered it.
–Nicolaus Copernicus

Only two things are infinite, the universe and human stupidity, and I'm not sure about the former.
–Albert Einstein

CHAPTER I: INFINITE TIME

As philosophers of science often point out, even though the concept of time has been studied since antiquity, it still is not well understood, and its definition is not well established. Therefore, I have to carefully state at the outset of this chapter what I mean by "time." In this chapter, I will use the common concept of time as defined by human clocks, a concept that I call "human time." (In chapter II, I will discuss the concept of "physical time.") A heartbeat, a pendulum, and the phases of the moon are examples of human clocks. Human time is meant to correspond to the common concept of time that all people understand. The concept of time that everyone understands, perhaps without explicitly realizing it, is all one needs to understand and appreciate the contents of this chapter.

1 Fundamental Premises of Standard Cosmological Model

The central assumption underlying the standard cosmological model is that the universe is expanding, which implies that the universe was created at a certain time in the past—that is, at the time the expansion started. This means that past time is finite. The standard model was developed based on a solution of the equations of general relativity. It is currently accepted, as a physical reality, by the majority of physicists. Yet, that reality is questioned by some highly accomplished physicists. For instance, Jean-Claude Pecker writes, "... another universe than that of the astrophysicists is offered by mathematicians who conceive all sorts of objects and geometries. But there is too great a temptation to consider these constructions 'real', as soon as they are plausible."[7]

In this book, I will discuss the cosmology of a universe that is nonexpanding on a very large time scale, which contradicts the standard cosmological model. Such a discussion, of course, would make no sense if the expansion of the universe had already been confirmed. Therefore, I must first show that no such confirmation is available so that putting forward a model of a nonexpanding universe is not only justifiable but, in fact, highly desirable from the perspective of progress in science. I want to emphasize that the discussion presented in the remainder of this section is not intended in any way to be a critique of the standard (big-bang) cosmological model. That model may (or may not) be physically appropriate. I merely want to demonstrate that the fundamental premises of the standard model remain unconfirmed.

Let me first put the widespread acceptance of the standard model into the perspective of scientific progress. As a result of the way normal science is done, potentially feasible cosmologies other than the standard model are not usually presented to the general public (nonspecialists) or to undergraduates in physics, as the information on cosmology offered in scientific papers, university textbooks, magazines, television shows or popular-science books is largely restricted to discussions of the standard cosmological model. And most of the upcoming specialists will have little chance to ever think of other cosmologies after being told at their first course in cosmology, "That our universe is presently expanding is established without doubt," which is the opening statement in a recently published introductory textbook.[8] That is the way of normal science. In his famous essay, Thomas Kuhn explained how normal science is done.[9] The central proposition in Kuhn's explanation concerns why and when scientific PARADIGMS change. Perhaps his most striking example of a change in a paradigm is the change from the Ptolemaic to the Copernican cosmology. The key reason for that change was the system of EPICYCLES incorporated into the Ptolemaic cosmology, which was invented with the purpose of explaining the observed motions of the sun, the moon, and the planets. As the astronomic observations became more accurate with time, that system of epicycles, a classic example of an ad hoc hypothesis, became so complex that it lost its credibility. As a result, scientists realized the need to research other cosmologies.

While there are a number of ad hoc hypotheses incorporated in the standard cosmological model, perhaps the most recent one, the existence of dark energy, signals the need for a possible paradigm change (as a minimum, the need to consider other cosmologies). Dark energy has a peculiar property of exerting a negative pressure on space, which causes its accelerated expansion. The accelerated expansion of space was concluded in 1998 from high-z SUPERNOVA observations. (See the entry "HIGH-Z" in appendix A for a brief explanation of those observations.) What is of importance to this discussion is the fact that while dark energy has been conjectured to exist, its existence remains entirely unconfirmed. From the perspective of introducing an ad hoc hypothesis to uphold a cosmological paradigm, the existence of dark energy appears to be an idea reminiscent of the epicycles system introduced in Ptolemy's cosmology. Until the form of dark energy is identified and its existence as a physical reality is confirmed, the standard cosmological model will have to be seen as an unconfirmed model of the universe.

With specific reference to physics and astronomy, the way normal science is done has been explained in a series of essays published by some accomplished scientists and philosophers.[10] Those essays often put an emphasis on the journal-imposed restrictions on publishing the results of any out-of-the-mainstream research. Defying those restrictions, Martin Lopez-Corredoira discusses cosmological models other than the standard model, including some nonexpanding universe models.[11] His paper suggests that the concept of a nonexpanding universe cannot be disregarded at this stage of cosmological research, which means that the current cosmological paradigm is not carved in stone. That supports the need to consider the leading (nonexpanding-universe) conception put forward in this book.

Similarly, the expansion of space itself has never been confirmed. Within the framework of the standard cosmological model, the expansion of space means that galaxies recede from one another at speeds that are proportional to the distances between them. In this respect, nonspecialists and undergraduates in physics and cosmology are misled by the statements of some of the most proficient physicists and cosmologists. For instance, in reference to Edwin Hubble's famous 1929 paper,[12] the following statements have been made:

Roger Penrose, "... it became clear, from Edwin Hubble's observations in 1929, that the universe is expanding ..."[13]

Paul Davies: "With the discovery by Edwin Hubble and other astronomers that the universe is indeed expanding."[14]

Stephen Hawking: "... in 1929, Edwin Hubble made the landmark observation that wherever you look, distant galaxies are moving rapidly away from us. In other words, the universe is expanding."[15]

Frank Wilczek: "... as the evidence for the expansion of the universe firmed up in the late 1920s, mainly through the work of Edwin Hubble."[16]

Brian Greene: "In 1929, Edwin Hubble ... found that the couple of dozen galaxies he could detect were all rushing away."[17] (Green is incorrect in his reference to Hubble's findings. Hubble found that out of twenty-four examined NEBULAE only nineteen rather than "all" were rushing away. He showed that the remaining five galaxies were actually rushing toward the earth. The crucial importance of this finding will be discussed in section 44.)

Nearly identical statements have been made by other prominent scientists and organizations, including Michael Rowan-Robinson,[18] Alan Guth,[19] Lawrence M. Krauss,[20] Michio Kaku,[21] Bernard F. Schutz,[22] Anthony Zee,[23] NASA,[24] Peter Coles,[25] and many others. These sorts of statements have also been made, either explicitly or implicitly, in hundreds of research papers on various aspects of the standard model. However, neither of those statements correctly reflects Hubble's discovery. In this regard, I am not sure how carefully any of the above authors have read Hubble's 1929 paper.[d] In that paper, which is only four pages long, Hubble did not claim to have found, discovered, evidenced, or observed that galaxies are receding from one another. He merely *assumed* that the observed cosmological redshifts are due entirely to the Doppler shift, which is an unconfirmed assumption originally made by Vesto Slipher[26] about sixteen years before Hubble's publication. For Hubble to discover that the universe is expanding, he would have had to measure the velocities of the galaxies and then arrive at a *radial velocity–redshift* relationship that would prove the expansion of the universe. He did not measure velocities. He estimated the distances to a number of galaxies (primarily, to the so-called Cepheid nebulae) and discovered the famous *galaxy distance–redshift* relationship. Hubble called it "the law of red-shifts".[27]

In many later publications Hubble emphasized that the receding galaxies surmise is just one possible interpretation of the observed redshifts. For instance, in his 1937 book Hubble wrote, "The familiar

---

[d] Such a perfunctory reading of a paper is not unusual when it comes to references made in support of the big-bang model. For instance, Hawking in *A Brief History of Time* and, in their Nobel Prize lectures, John C. Mather and George Smoot claimed that Ralph Alpher, Hans Bethe, and George Gamow predicted the existence of the CMB in their 1948 paper, *The Origin of Chemical Elements*, Physical Review **73** 803 (1948). (See P. James Peebles's *Discovery of the hot Big Bang: What happened in 1948*, arXiv:1310.2146v2 [2013].) In fact, the authors of that paper (which is just one-page long) made no such prediction, the word 'radiation' was never mentioned, and no reference of any kind was made to the temperature of radiation today or ever.

interpretation of red-shifts [i.e., red-shifts are entirely due to the Doppler shift] seems to imply a strange and dubious universe, very young and very small. On the other hand, the plausible and, in a sense, familiar conception of a universe extending indefinitely in space and time, a universe vastly greater than the observable region, seems to imply that red-shifts are not primarily velocity-shifts."[28] Hubble never changed his mind. In 1953 he wrote in the last of his publications: "The high density [the mean mass density in the universe] suggests that the expanding models are a forced interpretation of the observational results."[29] A number of scientists (e.g., Andre K. Assis et al.[30]) draw attention to the fact that the Doppler-shift interpretation of redshifts was, in Hubble's view, a mere possibility rather than a confirmed physical phenomenon.[e] A thorough and, as far as I can see it, unbiased review of possible problems with the big-bang interpretation of cosmological redshifts and other astronomic observations that are used to support the standard cosmological model is provided by Lopez-Corredoira.[31] While I have no interest in arguing here the weak points of the standard model, I believe that anyone curious about its confirmed or unconfirmed physical reality would benefit from reading Lopez-Corredoira's paper.

It needs to be said that in rare instances, leading physicists point to the correct meaning of Hubble's findings. For instance, Nobel Prize laureate, Steven Weinberg, wrote,[32] "However, it was discovered in the decade 1910–20 by Vesto Melvin Slipher of the Lowell Observatory that the spectral lines of many nebulae are shifted slightly to the red or blue. These shifts were immediately *interpreted* as due to a Doppler effect, indicating that the nebulae are moving away from or towards the earth…This *interpretation* became generally accepted after 1929, when Hubble announced that he had discovered that the red shifts of galaxies increase roughly in proportion to the distance from us [italics added]."

---

[e] Note that Slipher, the father of the Doppler interpretation of cosmological redshifts, clearly understood that it was just one possible interpretation. In his own words, "The magnitude of this velocity, which is the greatest hitherto observed, raises the question whether the velocity-like displacement might not be due to some other cause, but I believe we have at the present no other interpretation for it. Hence we may conclude that the Andromeda Nebula is approaching the solar system with a velocity of about 300 kilometers per second."

Indeed, Hubble's discovery of the relationship between distances to galaxies and galaxy redshifts was one of the greatest discoveries in astronomy. But the receding-galaxies conjecture merely remains an unconfirmed interpretation of the galaxy redshift observations. Since that interpretation of redshifts and the resulting consequences of it present the most outstanding argument against the idea of the infinite universe put forward in this book, I will discuss some directly relevant background information in the remainder of this section. (In section 44, a further discussion in this regard and another interpretation of cosmological redshifts will be presented.)

The dearth of proof that would confirm the recession of galaxies—that is, the conception of an expanding universe—is particularly well illustrated by the statement of Smithsonian Astrophysical Observatory prepared for the NASA web page: "Galaxies are so large, and so far away, that you could never see them move just by looking—even if you looked for a whole lifetime through the most powerful telescope!"[33] That statement applies to the motions of galaxies in general and to the radial motions of galaxies in particular. In other words, the premise of receding galaxies cannot be practically confirmed based on direct evidence. It can only be assumed based on the interpretation of cosmological redshifts as shifts that are due entirely to the Doppler-shift effect. Note that there are direct methods of measuring radial velocities of cosmic objects; however, those can be used to measure the motions of nearby stars and planets rather than the motions of faraway galaxies (see, e.g., Dainis Dravins et al.[34]).

Other interpretations of cosmological redshifts are possible and have been proposed. A few months after the 1929 Hubble paper was published, another distinguished astronomer, Fred Zwicky, proposed (in the same journal) an alternate explanation of cosmological redshifts that has become known as the "tired light hypothesis."[35] Hubble later considered that hypothesis to be an alternative on par with the Doppler-shift hypothesis. (After Zwicky put forward his hypothesis, specific tired light explanations were proposed by a number of authors, including P. A. Violette,[36] J. C. Pecker et al.,[37] J. C. Pecker and J. P. Vigier,[38] Michael Harney,[39] Jose F. G. Julia,[40] Amitabha Ghosh,[41] and others.) The key idea of the tired light hypothesis is that light beams, which comprise photons, lose energy as a result of interactions with some particles or media while traveling over the vast distances of the universe, which results in the generation of cosmological redshifts. However, since any tired light hypothesis contradicts the standard model, those hypotheses have been largely ignored by mainstream scientists, who have chosen to believe that the Doppler shift is the only cause of the observed galaxy redshifts.

I believe that presenting the standard Doppler-shift explanation of cosmological redshifts with reference to the sound-wave analogy, which is normally done in discussions of the standard cosmological model, impedes our ability to arrive at a more rational interpretation of the observed galaxy redshifts. In this regard, it seems to me that it is better to recognize that the beams of light comprise particles (photons) rather than waves.[f] This empirical fact was stressed by Richard Feynman, who wrote in 1985: "I want to emphasize that light comes in this form—particles. It is very important to know that light behaves like particles, especially for those of you who had gone to school, where you were probably told something about light behaving like waves. I am telling you the way it does—behave like particles."[42] The fact that light comprises particles and not waves has been known since Max Planck's 1900 presentation on radiation quanta[43] (which were later called "photons" by Albert Einstein) and Einstein's paper on the photoelectric effect published in 1905.[44] A particle-based interpretation of cosmological redshifts will be discussed in section 44.

The expansion of space according to the standard cosmological model requires geometric space-time, the concept introduced by Hermann Minkowski in 1908,[45] to be a physical reality. That is because space-time is the backbone of general relativity (Einstein's theory of gravitation), which, in turn, is the foundation of the standard model. However, confirming that space-time is a physical reality would require detecting it regardless of (i.e., separate from) any physical theory or model. I want to emphasize that space-time has never been observed or

---

[f] The wavelike character of a particle is demonstrated by the effectiveness of a wave function in describing the particle's behavior. As far as I know, no analogy has ever been found between the wave function of a photon (a mathematical concept) and a sound wave, the pitch of which can be explained by the physics of the Doppler-shift effect. Note that any periodic phenomenon, such as a beam of light comprising photons, can be described using wave mathematics.

detected. All that has been shown over the last hundred years is that curved space-time allows for a highly accurate mathematical description of gravitational interactions. It is also worth noting that the geometric interpretation of gravitational interactions, which is the basis of general relativity, is not compatible with other physical interactions such as ELECTROMAGNETIC, STRONG, AND WEAK NUCLEAR FORCES (see, e.g., Weinberg[46]).

In summary, the fundamental premises of the standard cosmological model—space-time as a physical reality, the big-bang "explosion," the creation of mass-energy out of nothing, the receding motions of galaxies, the existence of dark energy, and the early inflation of the universe—remain unconfirmed. While that does not prove that the standard cosmological model is erroneous, it certainly encourages one to study alternate cosmologies that could be more amenable to verification. Many scientists identified problems with the key premises of the standard cosmology that have not been confirmed (e.g., Tom Van Flandern,[47] William C. Mitchell,[48] Lopez-Corredoira,[49] Michael J. Disney,[50] and Yurij Baryshev[51]).

2 Infinite-Time Hypothesis

By the early 1970s, the majority of physicists and cosmologists accepted the standard (big-bang) cosmological model. A fundamental assumption underlying that model is that time, space, mass, energy, and the laws of physics were all created out of nothing at the very moment of the big bang. In particular, the model assumes that human time had a beginning, which means that past time is finite and will, or will not, have an end. In this book, the perfect cosmological principle is assumed to hold true. This implies that time is infinite, a notion that I will call the *infinite-time hypothesis*. In other words, I assume that time had no beginning and will have no end. There is no basis for that hypothesis other than aesthetics and the fact that any cosmology based on it can be consistent with the mass-energy and other conservation laws.

The widely accepted idea of finite time comes from the belief that the universe is expanding. As I explained in the preceding section, the expansion of the universe is not supported by any direct evidence. Beliefs embedded in some religions aside, I am not aware of any arguments in favor of the premise of finite time other than the unconfirmed Doppler-shift interpretation of cosmological redshifts.

Aristotle believed that the universe is infinite in time. In more recent times, Giordano Bruno—in his 1584 essay titled "On the Infinite Universe and Worlds"— famously proposed the conception of the universe that is infinite in time and space. Bruno's ideas were abandoned following Georges Lemaitre's proposal put forward in 1931, which involved a beginning of time and a finite-in-space, expanding universe (the original big-bang model). Bruno's conception of the universe was generally abandoned in favor of the big-bang model because the latter incorporated the assumed expansion of the universe, and not because of any disagreements with experiments/observations. As the cosmological model put forward in this book is in general agreement with Bruno's conception, the primary objections raised with respect to that conception need to be addressed. I will discuss those in the remainder of this section. It will be important to keep in mind that the spatial infinity of the universe is inherently related to infinite time, as I will discuss later, in section 3.

First, let me note that the premise of the universe that is infinite in space and time has not been entirely abandoned. It has been proposed as an alternative to the big-bang model by some physicists (e.g., by Assis[52]). Note also that the premise of infinite time is associated with the so-called cyclic cosmological models, in which the universe is subject to an endless sequence of the big-bang–big-crunch cycles as proposed, for instance, by Paul Steinhardt and Neil Turok.[53] Infinite time in such models, however, is not strictly infinite in physical sense since no physical history of the universe can be conveyed between cycles: all information, matter, energy, as well as the laws of nature would be annihilated at the time of each big crunch.

In support of general relativity, Einstein[54] presented a proof (hereinafter referred to as EINSTEIN'S PROOF) designed to show that the Newtonian concept of gravity, in conjunction with spatially-infinite universe, had to be incorrect. For that proof to hold, however, Einstein had to assume the existence of two infinities: infinite space and infinite range of gravitational interactions. From the perspective of the GPU cosmology, it is important to realize that for a spatially-infinite universe, Einstein's proof shows that the range of gravitational interactions ("the range of gravity") cannot be infinite and thus has to be finite. Put otherwise, if the range of gravity is finite, Einstein's proof has no meaning, and a spatially-infinite universe is possible. (As stated earlier in this section, the infinite-time hypothesis implies that the universe is also spatially infinite.) Alan Guth[55] and Stephen Hawking[56] point to another version of Einstein's proof. Again, for the Guth-Hawking argument to hold, an infinite range of gravity has to be assumed.

The problem with the infinite range of gravity assumed by Einstein, Guth, Hawking, and nearly all other physicists since Newton, who was first to propose an infinite range of gravity, is that it leads to the following gravity paradox: *gravitational interactions have an infinite range even though masses, which are, in some way, the measure of gravitational effect, are finite*. Since the mass of an object cannot be infinite, the only way to resolve this paradox is to postulate a finite range of gravity. It follows that both the infinite-time hypothesis via Einstein's proof and the only possible resolution of the gravity paradox suggest that the range of gravity is finite.

In the same book, Hawking presents another argument against the spatially-infinite universe—the well-known Olbers's paradox: in a spatially-infinite universe, "the whole sky would be as bright as the sun." That argument, however, requires one to make the assumption that all stars emit radiation in the VISIBLE-LIGHT frequency range. Thus, Hawking's argument is not rational as an argument against the spatially-infinite universe unless one pre-assumes that the universe has existed for a very short period of time so that stars have not had enough time to cool down to temperatures lower than those corresponding to the visible range of RADIATION FREQUENCIES. (Olbers's paradox is further discussed in the last paragraph of section 12). Hawking also refers to Olbers's counter-argument, which was, "...that the light from distant stars would be dimmed by absorption by intervening matter." He further states that "if that happened, the intervening matter would eventually heat up until it glowed as brightly as the stars. The only way of avoiding the conclusion that the whole of the night sky should be as bright as the surface of the sun would be to assume that the stars had not been shining forever but had turned on at some finite time in the past." Herein, Hawking again pre-assumes that stars can emit radiation only in the visible-light frequency range. The fact is, however, that some stars might have been shining for a very long time and could be emitting radiation in, say, the microwave frequency range with the intervening matter *not* being heated up to any significant extent, just like in the case of the cosmic microwave background (CMB), which, without doubt, does not heat up the intervening matter to the point that it glows as brightly as visible stars.

3 Cosmological Consequences of Infinite Time

The sun is shining. This very first human observation is central to the understanding of the physics of the infinite-in-time universe. Neither the sun nor any other star can ever stop shining (i.e., stop emitting energy). Otherwise, given infinite time, there would be an infinite number of stars that no longer emit energy. That a star cannot stop emitting energy is also required by the second law of thermodynamics (SLT), as I will discuss in section 31. Since stars never stop emitting energy, each star must eventually die given infinite time—that is, it must disappear by emitting its entire energy (mass). That also means that, on average, for any star that dies, another star must be born. Otherwise, given infinite time, all stars would have already died, which, as we well know, is not the case.

At least some stars must be born out of the electromagnetic radiation emitted by other stars. Otherwise, given infinite time, the matter available for the formation of stars that comprises COSMIC GAS AND DUST would be exhausted. As a result, there should be a deficit in the radiation of stars: stars should emit more radiation energy than they absorb from the emissions of other stars such that there is energy left for the creation of matter necessary for the formation of new stars.

A star that is dying by emitting its energy (its mass-energy is being dispersed) causes an increase of ENTROPY in the universe because the disorder in the universe increases. A star that is being born out of radiation causes a decrease in the entropy of the universe, since the disorder in the universe decreases. The total entropy in a representative volume of the universe (RVU) cannot increase or decrease on a large time scale—that is, it must remain constant on such a scale, which is consistent with the conclusion that for each star that dies, another star has to be born. Otherwise, given the infinite time, the amount of entropy would be infinite or zero, respectively, which is not the case.

The universe that is infinite in time must also be spatially infinite. That is for given infinite time, all masses in a spatially finite universe would have already collapsed to its gravity center, which, of course, is not the case.[g] (This is Newton's original argument stated in support of a spatially-infinite universe.) A spatially-infinite universe cannot expand, as there is nothing to expand into.[h] Nor can such a universe evolve on a large-time scale, since any evolution requires a beginning.

Given infinite time, there should exist a relatively large number of "invisible" stars/galaxies that emit radiation at very low BLACKBODY temperatures. This means that photons emitted from those stars/galaxies would be expected to comprise very low energies.[i] This is for, consistent with the Stefan-Boltzmann law, the rate of radiation emission decreases with the blackbody temperature of the emitting body to the fourth power. (See equation (27) in section 32, which states the Stefan-Boltzmann law.) What this means is that the number of galaxies emitting radiation at temperatures, say, between 5 K and 10 K would be expected to be much greater than the number of galaxies emitting radiation at temperatures, say, between 1995 K and 2000 K. That is for the rate of radiation emission from a galaxy having a blackbody temperature of 5 K would be roughly 2.5×10^{10} times lower than that from a galaxy having a

---

[g] The idea of spatially-infinite universe is very old. According to Thomas Kuhn: "In the fifth century B.C., the Greek philosophers, Leucippus and Democritus, visualized the universe as an infinite empty space, populated by an infinite number of minute indivisible particles or atoms moving in all directions. The earth was one of infinite, essentially similar heavenly bodies formed by the chance aggregate of atoms". The idea of spatially-infinite universe has survived until the big-bang model was proposed in the early 1930s.

[h] In the standard cosmological model, the "nothing-to-expand-into" inference is avoided based on the assertion that space-time exists as a physical reality, which allows for a universe that is finite in size and has no boundaries. In the universe that has no boundaries, there is no space outside of the universe, and its expansion does not require something to expand into. In the GPU cosmology, no assumption that space-time is a physical reality is made, whence a finite, expanding universe would have to have a boundary. Then it would need something (space) to expand into. The question of whether space-time is or isn't a physical reality will be discussed in section 18.

[i] In terms of radiation emission, a star acts approximately as a blackbody. The rate of radiation emission from a blackbody depends on its surface temperature and surface area.

blackbody temperature of 1995 K, assuming that both galaxies have the same "surface area." That means that a 1995-K galaxy is expected to lose its mass-energy much faster than a 5-K galaxy. Consequently, it is also expected that the vast majority of galaxies (more accurately, the stars and cosmic dust/gas that form those galaxies) should emit radiation at very low temperatures—that is, at very low radiation frequencies. If that is so, we should be able to detect the very low-frequency radiation, and seemingly, we do. In a spatially-infinite universe, it would have to be the CMB radiation, unless there is another low-frequency radiation coming from all directions in the universe that has not been detected yet.

In a spatially-infinite universe, the line of sight of any observer has to be intercepted by a star, which emits radiation. (That is a requirement of the cosmological principle.) All those stars form a "star horizon." In other words, any observer in the universe is located inside a giant box that has highly rugged walls. The walls form a continuous enclosure that emits thermal radiation toward each observer. The great majority of the walls comprise very old and very cold stars. Those stars are invisible to human eye. Only a very small fraction of all stars would be expected to emit radiation in the frequency range of the visible-light and higher. The existence of invisible stars/galaxies means that the average mass-energy density in the universe would be expected to be much higher than the density that could be estimated from accounting for visible galaxies only.

In summary, if the infinite-time hypothesis holds, then the universe must be spatially infinite, nonexpanding, and nonevolving on a large time scale; the average mass-energy density in the universe has to be much higher than the density estimated from accounting for visible galaxies only; and the entropy in the universe must be constant on a large time and space scale. Furthermore, the source of the CMB-like radiation is expected to be very low-temperature galaxies; stars must be born out of radiation and die by emitting out their entire energies, whence stars emit more energy than they absorb; and the range of gravity must be finite.

The foregoing discussion allows for a high-level summary of the GPU cosmology, which compares with the standard cosmological model as follows:

- The principal features of GPU cosmological model include infinite time (there was no act of creation), infinite space, nonexpanding universe, finite range of gravity, conservation laws satisfied with no exceptions; and the perfect cosmological principle holds.
- The principal features of standard model cosmology include finite past time (the big bang was the act of creation), finite or infinite space-time, expanding universe, infinite range of gravity, violation of conservation laws at the big bang; and the cosmological principle (as opposed to the *perfect* cosmological principle) holds.

CHAPTER II: DEHUMANIZATION OF NATURE

To appreciate the cosmology of the infinite-in-time universe, it will be sufficient to read sections 4, 6, 7, the first paragraph of section 9, and section 10. To appreciate all discussions presented in chapters III and IV, it will be necessary to read sections 4 and 6 through 11. To fully appreciate the simplicity of both nature and the universe, the entirety of chapter II should be read.

4 Background

Similar to the ancient Greeks, who humanized their gods; modern physicists have humanized nature (physics). Most physicists believe that nature knows and applies advanced mathematics to solve the equations invented to state physical laws, for instance, the equations of Maxwell's electromagnetic theory or Einstein's field equations. They also believe that nature is capable of measuring physical quantities (e.g., the quantity of charge or mass), space, and time. Furthermore, physicists believe that nature makes decisions concerning the amounts of physical interactions, for instance, concerning the strength of physical interaction (the strength of force) depending on the results of such measurements. We know that solving equations and performing other mathematical chores, measuring quantities, making decisions, and similar abilities are attributes of the human mind. We cannot know, however, if nature knows mathematics, measures quantities, or makes decisions, just like the ancient Greeks could not possibly know if Aphrodite really was very beautiful or if Zeus actually cheated on Hera. By assuming that nature has those abilities, physicists have humanized her.

In this chapter, I will dehumanize nature (physics) by denying her the ability to solve equations, measure quantities, or make decisions, and examine some of the resulting implications. I will do that by putting forward a fundamental-law hypothesis that will suggest classical physics at the fundamental level to be extraordinarily simple. Dehumanized physics will allow for the rational explanation of gravitation and the constancy of the speed of light, both of which are key elements of the GPU theory.

5 Some Crucial Issues in the Philosophy of Physics

The nature of space and time has been debated since antiquity. Yet the nature of space and, particularly, the nature of time still remain mysterious. Bradley Dowden presents a detailed account of the various concepts of time put forward over the centuries that shows that we are not much closer to comprehending the nature of time than the ancients were.[57] Perhaps the most comprehensive discussion of time from the perspective of modern cosmology is that of Sean Carroll.[58] With specific reference to physics and cosmology, and without attempting to define time or space, I will suggest later (in section 13) that the difficulties with comprehending the concepts of space and time arise because the fundamental laws of nature do not incorporate those concepts. Most modern studies of space and time focus on specific aspects of the current physical theories, primarily the quantum theory and the theory of general relativity (e.g., Arthur Eddington,[59] Craig Callender and Carl Hoefer,[60] Holger Lyre,[61] Craig Callender[62].) In contrast, in the discussion of space and time presented in section 13, I will focus on the relation of space and time to the fundamental laws of physics and to physical phenomena in general, rather than to the specifics of physical theories.

The role of mathematics in physics has been debated since antiquity as well. The interest in this role has intensified over the last half century, as a number of renowned physicists have explicitly argued that mathematics is somehow built into the fabric of nature. According to Mark Colyvan, philosophers are less interested in that notion.[63] That could be because, according to Hawking's belief, "In the nineteenth and twentieth centuries, science became too technical and mathematical for philosophers."[64] (Later, Hawking went further, declaring that "these are questions for philosophy, but philosophy is dead."[65]) While Hawking's belief may or may not be correct in general, it certainly is discourteous toward many philosophers of science, some of whom are clearly at ease with mathematics. Regardless, the difference between mathematical (or "applied") and theoretical physics has to be first understood and appreciated. Philosophers do not necessarily have to use mathematics to advance the understanding of nature—that is, to advance theoretical physics. I will demonstrate in this book that to advance theoretical physics, philosophical reasoning can be as powerful as mathematics. The flagship demonstration will be showing that the fundamental laws of physics can rationally be identified from a philosophical hypothesis put forward based on experimental physics, with no mathematics involved. Note that even some of the most eminent physicists do not realize the potential for contributions to physics from philosophers of science. For instance, Paul Dirac stated that "The physicist, in his study of natural phenomena, has two methods of making progress: (1) the method of experiment and observation, and (2) the method of mathematical reasoning."[66] This shows that Dirac implicitly discredited possible contributions of the philosophers of science to the advancement of theoretical physics.

I believe that the philosophy of science should be a key branch of physics. Consequently, in discussing the fundamental concepts of physics, I will pay as much attention to the views of philosophers of science as to the views of physicists. I also believe that the understanding of nature is not really necessary to advance applied (as opposite to theoretical) physics. On this point, one of the most gifted physicists, Richard Feynman, commented, "I think I can safely say that nobody understands quantum mechanics." If he was right, then I also am right in saying that there is no necessity to understand nature in order to advance applied physics. In accord with Feynman's comment, this can best be demonstrated by the development of the highly successful quantum theory, which I consider to be a branch of applied physics. It seems to me that relatively little effort has been made by scientists to gain a conceptual understanding of physics, whether at the quantum or the classical level. With respect to quantum physics, this seems excusable, as the human brain has developed affected by the macroscopic world, and, owing to the absence of direct experimental evidence, we probably do not know yet how to pursue the understanding of nature at the subatomic level. However, the relatively little effort that has been made to pursue a conceptual understanding of classical (macroscopic) physics seems to have no excuse.

6 Fundamental-Law Hypothesis

The following *fundamental-law hypothesis* is put forward: the fundamental laws of nature (physics) do exist independent of the human mind. In the context of this hypothesis, "fundamental" means that a fundamental law of physics (FLP) does not have underlying principles.

Consequently, a fundamental law of physics must not incorporate other FLPs. The fundamental-law hypothesis (FLH) is also meant to apply to matter and energy in the sense that fundamental particles (of matter and energy) do exist independent of the human mind.[j] That aspect of the FLH will be used in section 30, in which I will suggest that photons are fundamental particles. In summary, the FLH states that there are fundamental laws of nature and fundamental particles of matter and energy that exist independent of the human mind.[k]

Incorporating the phrase "independent of the human mind" in the FLH may seem naïve or trivial. Yet it is neither. That phrase stipulates

---

[j] To fully appreciate the following comment, basic knowledge of physics at the undergraduate level may be required: The concepts of fundamental laws and fundamental particles are closely related. Let me explain this relation using the example of strong nuclear force and quarks as the particle components of protons and neutrons. In the standard model of particle physics, the residual strong nuclear force, which binds protons and neutrons, is assumed to result from the execution of a physical law at the quark level. The question is whether both that law and quarks represent fundamental concepts that exist independent of the human mind. If quarks, as constituents of protons and neutrons, represent a mathematical concept introduced to describe the interactions between protons and neutrons (i.e., a concept that exists in the human mind only), then as far as we currently know, protons and neutrons are fundamental particles. As such, the physical law believed to be responsible for the generation of strong nuclear force (the gluon exchange) would not have the status of a fundamental law of nature. The fundamental law of nature underlying the interactions between protons and neutrons would be unknown. If, on the other hand, quarks are true physical (rather than mathematical) entities, then as far as we currently know, they are fundamental particles that exist independent of the human mind, while protons and neutrons cannot be seen as fundamental particles since they have underlying constituents, principles, and mechanisms (quarks, gluons, and their interactions). In this case, the gluon exchange as a law of physics that underlies the generation of strong nuclear force would be, as far as we currently know, a fundamental law of nature.

[k] With respect to matter and energy, the particles that satisfy the FLH include the electron, the positron, the photon, and perhaps some other particles such as neutrinos. Those particles do not decay. Other subatomic particles, such as the proton and the neutron, cannot be considered fundamental as the transformation of a proton into a neutron, or a neutron into a proton, involves the emission of other particles (electron, neutrino and their anti-particles), while the FLH implies that a fundamental particle cannot contain other particles or mechanisms that would lead to the creation of other particles.

that the fundamental laws of nature do exist regardless if we (or, more accurately, our consciousnesses) are there or not. It is a basic question of the philosophy of science that goes back to George Berkeley (1685–1753), who believed that neither the laws of nature nor physical entities, such as an apple or a star, exist except as ideas perceived by the human mind. While Berkeley's philosophy could be seen as absurd from the perspective of today's scientific method, it is still out there and in the running. This is well illustrated by the statements of two most prominent physicists, both Nobel Prize laureates. In a 1931 interview, Max Planck was asked, "Do you think that consciousness can be explained in terms of matter and its laws?" His response was, "No. I regard consciousness as fundamental. I regard matter as derivative from consciousness. We cannot get behind consciousness. Everything that we talk about, everything that we regard as existing, postulates consciousness."[67] In a 1961 paper, Eugene Wigner expressed a similar opinion, writing, "It will remain remarkable, in whatever way our future concepts may develop, that the very study of the external world led to the conclusion that the content of consciousness is an ultimate reality."[68] Those two quotes show how important (and daring) the FLH is.

The FLH requires that FLPs do not have underlying mechanisms, as all mechanisms operate according to some principles. It also requires that an FLP always governs physical phenomena in the same way—that is, it must be absolute regardless of circumstances such as the time, location, or specific physical setting. Otherwise, it would have to incorporate some underlying principles based on which different actions would be imposed in response to different circumstances. The statement that an FLP must be absolute may also be rephrased as: no decision making can be incorporated in an FLP as any decision making would require the FLP to incorporate some underlying principles. From similar reasoning, an FLP cannot incorporate a memory of any kind. Furthermore, an FLP must be nonquantitative as any quantitative action would require a measurement of a quantity and making a decision on the amount of action to be applied (a concern of particular significance that will be discussed in section 10). An FLP must also be nonmathematical since it cannot incorporate any, including mathematical, principles. Put otherwise, the FLH implies that nature does not do any mathematics. The above statements define strict constraints on FLPs, which are a direct consequence of the statement of the FLH.

Nature's inability to do mathematics appears to be the most controversial implication of the FLH, whence it calls for an additional explanation. Let me use general relativity as an example to make clear what I mean by nature's inability to do mathematics. General relativity is Einstein's theory of gravitation, the understanding of which is not essential here. What needs to be known for the purpose of this discussion is that according to general relativity, all objects subject to gravitational interactions move along some precisely determined pathways in curved space-time. (If you don't know what "curved space-time" is in physical terms, don't be hard on yourself. Chances are good that nobody knows.) Those pathways are called geodesics. If general relativity is a true law of nature, then nature has to do a lot of mathematics. In each instant, she has to measure the mass of each gravitationally interacting object in the universe, calculate a geodesic for each of those objects—for instance, for each star, planet, or apple falling from a tree—and then force somehow each of those objects to move along the calculated trajectory. (Keep in mind that a geodesic comprises neither matter nor energy so, in itself, it cannot interact with objects) If, instead, general relativity is, as I believe, a mathematical description of gravitational interactions, then nature does not do any mathematics. In that case, each gravitationally interacting object follows a trajectory set by a nonmathematical, fundamental law of gravitation, conflicting (interfering) laws of nature, and/or interfering matter or energy, as I will discuss in section 10. Then, the curved space-time represents a mathematical instrument that exists in the physicist's mind only rather than a physical reality.

Let me now use kinetic energy as an example to explain what I mean by nature's inability to measure quantities or to make decisions. The nonrelativistic formula that determines the amount of kinetic energy (KE) of an object in motion is $KE = 0.5 \times m \times v^2$, where m and v are the mass and the speed of the object, respectively. Let the object be an apple falling from a tree. When the apple collides with the ground, a conflict between a law of gravitation and the earth's surface (interfering matter) arises. As a result, the kinetic energy of the apple is converted into the heat energy that is generated and dissipated in the collision. The amount of converted energy depends on the mass of the apple. The following question arises: does nature measure the speed and the mass of the apple and then do mathematics to determine how much heat energy will be generated in the collision? Now, if $KE = 0.5 \times m \times v^2$ were a true law of nature, nature would indeed be expected to calculate the magnitude of KE prior to the time of the collision and determine the amount of the heat energy to be generated. The FLH's response to this is, "No way, hombre." Nature does not have a clue as to what the amount of kinetic energy of the falling apple would be at any time. (In the first place, she does not even know what kinetic energy is.) The quantitative effect of the apple colliding with the earth depends on the amount of conflict. Apparently, the greater the mass or the speed of the apple, the more conflict there would be. The clever physicist, on the other hand, knows well how much energy will be generated. S/he simply measures or calculates the mass of the apple and its velocity, and then uses the mathematical formula to calculate the energy that will be released.

The FLH comprises three premises. First, there is a physical reality that exists independent of the human mind. Second, there exist laws of nature (physics) that are independent of the human mind, a fact that is suggested by the observed regularities of physical phenomena. Third, there is a bottom level with respect to the description of a physical phenomenon at which the governing law of physics is not underlain by any principles. Such a law represents a fundamental law of physics. Without a bottom level, the hierarchy of the laws of physics would be infinite, which is untenable. Thus, at the bottom level, the laws of physics represent fundamental principles of physics that are independent of one another, which is a strongly reductionist point of view. However, I want to emphasize that my view of reductionism is in disagreement with the concept of reductionism held by many physicists. (See, for instance, Gerard 't Hooft[69].) In my view, reductionism with respect to the laws in physics applies to the fundamental laws of physics and not to physical laws, which are invented by physicists in order to describe physical phenomena in quantitative terms, and which exist in the human mind only. Examples of physical laws are Kepler's laws of planetary motion and Coulomb's law of electricity. The difference between a physical law and a fundamental law of physics will be discussed in section 12.

At this point, it may be useful to look at some laws in physics that do not satisfy the FLH. As an example, consider Newton's law of gravitation $F = Gm_1m_2/d^2$, where m_1 and m_2 are the masses of two gravitationally interacting objects, d is the distance between the gravity centers of the two objects, and G is a constant. That law incorporates an unknown mechanism that determines the magnitude of the force of gravity, which depends on the two masses and distance d. Therefore, the execution of Newton's law would require making measurements and calculations. Consequently, that law cannot hold the status of an FLP. Another example of a well-known law in physics that does not satisfy the FLH is Fourier's law of heat flow, $q = k\nabla T$, where q is the heat flux density, k is the thermal conductivity of the conductor, and ∇T is the temperature GRADIENT. That law incorporates a mechanism, which comprises the molecular and the atomic motions underlying the flow of heat energy. It follows that Fourier's law cannot hold the status of an FLP. There are other laws in physics—such as the standard model of particle physics, the ideal gas law, Faraday's law, general relativity, Ampere's law, Hooke's law, and many other laws—that cannot hold the status of an FLP because they incorporate some underlying principles, mechanisms, and/or other laws.

It is worth noting that the principle of relativity, which is one of the two basic postulates underlying Einstein's SPECIAL RELATIVITY, is not only fully consistent with the FLH but can also be concluded directly from that hypothesis. The principle of relativity states that the laws of physics have the same form in all INERTIAL FRAMES OF REFERENCE. The same follows directly from the FLH: having an FLP that has different forms in different inertial frames would require the FLP to incorporate a decision-making mechanism and some principles such that decisions could be made to enforce different laws depending on the observer's inertial frame of reference. That would contravene the FLH.

The FLH presents a philosophical thought that, I believe, cannot be proven or disproven. It can only be believed or not believed, and its correctness can be judged based on its ability to explain physical phenomena. Some physical concepts discussed in this book are directly based on the ability of the FLH to provide such explanations.

7 Fundamental Laws of Physics

The following laws of physics do not have known underlying principles:

I	The second law of thermodynamics (the SLT): nature eradicates non-equilibrium between physical entities (between physically connected locations containing different amounts of a physical quantity).
II	The law of inertia: mass resists a change to its state of motion.
III	The law of gravitational interaction: two mass objects attract each other.
IV	The law of electrostatic interaction: two unlike charges attract each other.
V	The law of electrostatic interaction: two like charges repel each other.
VI	The momentum-conservation law: momentum cannot be created or destroyed.
VII	The mass-conservation law: mass cannot be created or destroyed.
VIII	The energy-conservation law: energy cannot be created or destroyed.
IX	The charge-conservation law: charge cannot be created or destroyed.
X	The gravity-conservation law: Gravity cannot be created or destroyed.

Each of those laws is an FLP in agreement with the requirements of the FLH, which served as a touchstone for their identification. All of the identified FLPs represent experimental laws. They have been discovered from experiments or observations. From the perspective of experimental evidence, their statements are considered to be complete. Herein, "complete" means that the principles stated by those laws are, as far as we know, exact and absolute—that is, no deviation from any of those laws has ever been observed. For instance, the principle of two like charges repelling each other is exact and absolute based on all of the experimental evidence we have.

To appreciate the "exact and absolute" premise, consider the assumptions underlying the theory of general relativity: the speed of gravitational interaction equals the speed of light, and the gravitational acceleration of object A moving toward object B does not depend on the mass of object A, which is a statement of the equivalence principle. Those two assumptions are still being tested (see, for instance, papers by

Tom Van Flandern[70] and T. H. Wagner et al.[71]). Therefore, based on experimental evidence, neither of those two premises can be declared to be exact and absolute at this time.[1]

Nevertheless, the statements of laws III, IV, or V might not be complete. This is because their implications might not be valid under all circumstances. As an example, if the range of gravitational interaction is finite, as implied by the infinite-time hypothesis (section 3), mass objects would not attract one another at some large separation distances, and the above statement of law III would be incomplete because it does not account for its restricted validity. Yet, as long as two mass objects remain sufficiently close to each other for the gravitational interaction to occur, law III can be declared to be absolute and exact based on all experimental evidence.

The SLT is often said to be valid only statistically, which is a claim that has apparently been inspired by Ludwig Boltzmann's statistical interpretation of that law. According to that interpretation, the SLT would not be absolute, whence it could not hold the status of an FLP. However, the premise that the SLT is valid only statistically appears to be unsubstantiated, as I will discuss in section 11.

Law X (termed "the gravity-conservation law") is new. It emerges as a result of a finite range of gravity, the mass-energy equivalence, and the energy-conservation law. I will discuss the gravity conservation law in detail in section 21. As stated previously, a finite range of gravity is implied by the infinite-time hypothesis.

The above list of FLPs is expected to be incomplete, particularly with respect to the fundamental laws that govern physical phenomena at the subatomic (quantum) level. In this regard, I note that while current quantum physics appears to provide for a highly accurate mathematical description of physical phenomena at the quantum level, the fundamental laws of physics at that level are not well understood. This suggestion is supported by Feynman's famous statement on the understanding of quantum theory (see Feynman's quote in section 5).

---

[1] Herein, the reference is made to the statement of the weak (Galileo's) equivalence principle. In developing general relativity, Einstein actually used the strong equivalence principle. The latter principle accounts also for the effect of GRAVITATIONAL BINDING ENERGY, which is negligible for ordinary (relatively small mass) objects.

The simplicity of the identified FLPs suggests that they are perhaps what Steven Weinberg had in mind when he pointed out to the physicist's quest "for a simple set of physical principles."[72] Richard Feynman expressed similar thought, "I have often made the hypothesis that ultimately physics will not require a mathematical statement, that in the end the machinery will be revealed, and the laws will turn out to be simple."[73] Note that like some physicists, many philosophers of science, including, for example, Joseph Melia,[74] emphasize the expectation for the laws of nature to be simple.

There is a rather surprising conclusion that can be drawn from the statements of laws I–X. Neither space nor time is incorporated into any of those laws. That suggests that any physical law that incorporates space and/or time would not be fundamental—that is, such a law would have some underlying principles, mechanisms, and/or other laws. It needs to be said that the above conclusion is drawn in respect of the identified FLPs only. The discovery of a new FLP that satisfies the FLH and incorporates the concept of space or time would make that conclusion incorrect. The idea that time should not be incorporated in the fundamental laws of physics has been endorsed by many scientists, as pointed out by Julian Barbour.[75]

In summary, the attributes of FLPs that have been *hypothesized* herein (i.e., concluded by virtue of the statement of the FLH) include: an FLP does not have underlying principles, and it exists independent of the human mind. Since FLPs cannot incorporate underlying principles, which is a requirement of the FLH, it follows that FLPs must have no underlying laws, mechanisms, decision-making or memory abilities and must be nonmathematical and absolute. The attribute that has been *concluded* from the examination of the identified FLPs is that FLPs do not incorporate space or time.

It needs to be said that one cannot really hope to ever understand the FLPs, particularly laws I–V, which I call the "action laws," as one cannot know what is out there behind those laws or why they exist. That lack of understanding of the FLPs is a consequence of the lack of principles underlying those laws. Consider, for instance, the statement of the law of electrostatic interaction, "two unlike charges attract each other." Why or how this attraction is enforced is not understood now and, most likely, never will be. According to the FLH, no principles underlying that law exist. This situation is unacceptable to physicists who develop applied physics. The principles underlying physical interactions have to be defined such that physical laws, which describe physical phenomena in quantitative terms, can be formulated. Those principles are invented by physicists. Consider, for instance, the standard model of particle physics. In that model, the forces acting between particles of matter are assumed to be carried by the exchange of gauge bosons (virtual particles). Virtual particles, however, are not known to exist in the physical sense. Most likely, those particles represent mathematical entities incorporated into the physical laws that describe the forces of nature. While such laws can describe physical phenomena with an extraordinary degree of accuracy, those particles appear to exist in the human mind only.

The conservation laws (laws VI–X), on the other hand, are perhaps easier to understand if they are restated as follows, which also is suggested in the next section: nature has no hidden sources of, or storage bins for, physical quantities.

8 Correct Statement of a Fundamental Law of Physics

A correct statement of a fundamental law of physics is vital to the understanding of nature. Consider, for instance, the energy-conservation law. Physicists often think of conservation laws as being mathematical. Feynman thought of the energy-conservation law as follows: "It [the energy-conservation law] states that there is a certain quantity, which we call energy, that does not change in the manifold changes which nature undergoes. That is a most abstract idea, because it is a mathematical principle; it says that there is a numerical quantity which does not change when something happens…It is just a strange fact that we can calculate some number and when we finish watching nature go through her tricks and calculate the number again, it is the same."[76] If that were a correct statement of the energy-conservation law, that law could not hold the status of an FLP because, according to the FLH, nature is unable to calculate or measure quantities. The most typical mathematical statement of the energy-conservation law is: "The total amount of energy in an isolated system remains constant." For that statement to be correct, a mechanism would have to be built into the energy-conservation law that would first identify an isolated system and then measure its internal energies as well as perform calculations to somehow ensure that the total energy remains constant. Such a law could not hold the status of an FLP. Moreover, that statement of the energy-conservation law permits the creation of energy at one location in a closed system while the same amount of energy is simultaneously destroyed at another location. That has never been observed. The statement of the energy-conservation law "energy cannot be created or destroyed" is free from such contradictions while it carries all of the required attributes of an FLP. There is a down-to-earth yet fully adequate way to think of conservation laws: nature has no hidden sources of, or storage bins for, physical entities, from which it follows that neither energy nor any other fundamental physical entity can be created or destroyed.

Note that a mathematical statement of a conservation law would not provide any explanation as to why such a law exists. On the other hand, a more philosophical statement of a conservation law (i.e., "nature has no hidden sources of, or storage bins for, physical entities") provides a common-sense explanation in this regard. From the perspective of the FLH, stating a physical law in mathematical terms is a humanization of nature, which appears to be appropriate for the purpose of advancing applied physics as long as such a humanization is explicitly realized. My perception of the difference between applied and theoretical physics is explained in footnote q in section 14.

Another example of the need for a correct statement of an FLP is the statement of the SLT (law I in section 7). There are many statements of that law, most of which refer to heat transfer and some associated constraints. As an example, Lord Kelvin's statement of the SLT is: "It is impossible, by means of inanimate material agency, to derive mechanical effect from any portion of matter by cooling it below the temperature of the coldest of the surrounding objects." These types of statements refer to selected consequences of the SLT rather than to the law itself. As another example: the fact that entropy in an isolated system must not decrease is a consequence of the SLT, not a law in itself. That is because nature does not know how to calculate the amount of entropy—that is, the amount of disorder—and has no means to prevent entropy from decreasing. The requirement for entropy not to decrease is a simple consequence of the fact that nature enforces the eradication of nonequilibrium.

Most of the statements of the SLT refer to heat. That does not come as a surprise since the SLT was discovered at a time when scientists were preoccupied with heat generation and heat-engine efficiencies. Such statements, however, cloud the picture, as the SLT is applicable to any nonequilibrium, whether or not heat energy is explicitly accounted for. I will show later that applying the SLT to mass nonequilibrium with no consideration of heat is imperative to the understanding of gravitation (section 25) and the constancy of the speed of light (section 31).

9 Action-at-a-Distance

Following the discussion presented in this section could be a demanding task. A reader who has little time to spare or who does not like challenges can skip this section. However, that reader will have to then accept that an ACTION-AT-A-DISTANCE mode of physical interaction between matter or between matter and energy is possible in a space that is empty of matter and energy, provided that that space is filled with some potential for physical interaction like, for example, gravitational potential.

Discussing action at a distance is important because the statements of laws III, IV, and V identified in section 7 all imply an action-at-a-distance mode of interaction, while such a mode is not considered appropriate by the majority of physicists. On the first look, each of those laws implies a nonlocal mode of interaction between two masses or two charges, as there is nothing in the statements of those laws that would imply the existence of a field—that is, there is no implication of a local mode of interaction. A field in physics comprises an infinite set of field values, each of which is characteristic of a point in continuous space or space-time.[m] Those values determine the potential strength of an interaction (the strength of a force). In the description of physical interactions, a field-mediated interaction is a concept alternative to an action-at-a-distance mode of interaction. It is based on the principle of locality, which states that all matter and energy can only be directly influenced by its immediate surroundings—that is, by a local field. Any law of physics that is based on the concept of a field—for example, the Lorentz force law or the Gauss law for gravity—has to include some

---

[m] The concept of a field in physics also includes the force carriers (virtual particles that can be interpreted as quantum field excitations), which are incorporated in the standard model of particle physics.

mathematical principles, whence such a law cannot hold the status of an FLP according to the FLH. As I will discuss in detail later, in section 19, the GPU theory put forward in this book does include the concept of a field. That field, however, does not comprise any values (numbers).

As pointed out above, the instantaneous action-at-a-distance mode of physical interaction has not been favored by the majority of physicists, particularly after Maxwell published his theory of electromagnetism based on the concept of electromagnetic field. The concept of a field as the mode of physical interaction has been dominant in the development of applied physics ever since.[n] Since some of the FLPs identified in section 7 imply an action-at-a-distance rather than a field mode of interaction, this implications require a commentary, which is provided in the following paragraphs.

The dispute concerning an action-at-a-distance mode versus a field mode of interaction, which started at about the time of Newton, has not been settled yet, as many physicists as well as philosophers of science still view an action-at-a-distance mode to be a viable alternative to the concept of a field. That was suggested in a number of presentations at the conference titled "Instantaneous Action at a Distance in Modern Physics: Pro and Contra,"[77] in which some authors (e.g., Jayant Narlikar[78]) presented rather persuasive arguments in favor of the action-at-a-distance mode of interaction. There were more "pro" than "contra" authors at that conference; however, to be fair, that could be because the "contra" physicists were less motivated to attend such a conference as the contra opinion is consistent with the current paradigm of physics. Note that some physicists argue that the concept of instantaneous action at a distance and the concept of the Faraday-Maxwell field actually coexist.[79] Of interest also is that the action-at-a-distance mode of interaction has not been entirely abandoned in modern physics, as evidenced by the works of some most renowned physicists and astrophysicists, including Archibald Wheeler and Feynman[80] and Fred Hoyle and Narlikar.[81]

The standard argument of the critics of action at a distance is "how an object could know, looking at another object, how to move". Those critics, however, fail to provide an answer to the question of how an object does know an infinite number of field values supposedly existing

---

[n] The instantaneous action-at-distance mode of interaction is allowed in quantum physics, particularly with regard to the QUANTUM-ENTANGLEMENT phenomenon.

in space or space-time and then use mathematics to calculate the local field values required to determine how the object is to move.

Newton is normally portrayed as a strong opponent of the action-at-a-distance mode of interaction. In a 1692 letter to Richard Bentley, he famously wrote:[82]

> That gravity should be innate inherent & essential to matter so that one body may act upon another at a distance through a vacuum without the mediation of anything else by and through which their action or force may be conveyed from one to another is to me so great an absurdity that I That gravity should be innate inherent & essential to matter so that one body may act upon another at a distance through a vacuum without the mediation of anything else by and through which their action or force may be conveyed from one to another is to me so great an absurdity that I believe no man who has in philosophical matters any competent faculty of thinking can ever fall into it. Gravity must be caused by an agent acting constantly according to certain laws, but whether this agent is material or immaterial is a question I have left to the consideration of my readers.

In spite of Newtown's opposition to the action-at-a-distance mode of interaction, his law of gravitation does imply such a mode. This probably troubled Newton and may have led to another of his famous statements, "But hitherto I have not been able to discover the cause of those properties of gravity from phenomena, and I frame no hypotheses."[83] It is of interest to note that Newton's proclamation quoted above has recently been challenged by John Henry,[84] who argues that Newton's statement in the letter to Bentley did not reflect his true thoughts. Henry argues that Newton actually believed in an action-at-a-distance mode of interaction.

In 1916, Einstein followed Newton's stance in respect of action at a distance and went further in general relativity, in which he assumed that the geometry of Minkowski's space-time represents a gravitational field with a local geometric property (space-time curvature) that determines the strength of gravitational interaction.[85] Next to Newton, Einstein appears to be the most often-quoted opponent of action at a distance.

When referring to action at a distance in 1916, Einstein echoed Newton's thoughts: "We have come to regard action at a distance as process impossible without the intervention of some intermediary medium."[86] Einstein also expressed his disbelief in action at a distance with specific reference to the quantum theory: "I cannot seriously believe in it [in the description of quantum entanglement] because the theory cannot be reconciled with the idea that physics should represent a reality in time and space, free from spooky action at a distance."[87] (Today, Einstein's opposition to the quantum description of quantum entanglement is largely discredited, primarily because of its experimental confirmation.) Regardless, Einstein may have changed his mind with respect to action at a distance later. In 1954, after thirty-eight years of additional thinking, he wrote to his friend Michele Besso: "I consider it quite possible that physics cannot be based on the field concept, that is, on continuous structures."[88] In this regard, I note that the only known alternative to the field mode of physical interaction is the action-at-a-distance mode.

Ernst Mach, on the other hand, did not see anything wrong with action at a distance. He stated, "It is well known that action at a distance has caused difficulties to very eminent thinkers. 'A body can only act where it is'; therefore there is only pressure and impact, and no action at a distance. But where is a body? Is it only where we touch it? Let us invert the matter: a body is where it acts. A little space is taken for touching, a greater for hearing, and a still greater for seeing."[89] To interpret Mach's argument properly, one has to realize that *hearing* requires a contact by a medium (i.e., by air that carries sound waves) or, using Mach's terminology, it requires a "touch." Similarly, *seeing* does require a contact by a medium (electromagnetic radiation). Thus, seeing does require a touch as well. It follows that a *direct* touch is not the only alternative to the action-at-a-distance mode of interaction. The other alternative is an indirect touch—that is, a touch over a distance that uses an invisible mediating agent such as air or electromagnetic radiation. Mach's argument, therefore, does not really oppose the Newton-Einstein view, since his concept of "action at a distance" is based on the presence of mediating agents.

While I certainly agree with the Newton-Einstein view that a mediating agent is necessary to carry a physical interaction, I disagree with their concept of a vacuum in which no mediating agent exists. In particular, Einstein's explicit view was that if two objects were separated by a vacuum (i.e., by space with no matter or energy present), they could not possibly interact with each other, which indicated that a gravitational action at a distance was impossible. Let me explain my stance. To keep the discussion simple, I will refer to gravitational interactions only. Einstein's vacuum actually is a field, which is filled with the potential for gravitational interactions. If a grain of sand is put into Einstein's vacuum—that is, into any place in the universe that is free of matter and energy—it will be subject to gravitational interaction. (Einstein believed in infinite range of gravity, whence a mass-energy object would experience a "force of gravity" at any location in space or space-time.) Einstein did not recognize that a gravitational field represents a physical entity, as evidenced by the sand-grain test. As a physical entity, the gravitational field can be an agent that mediates gravitational interaction and, from this perspective, gravitational interaction would no longer be classified as an action-at-a-distance mode of interaction. It is worth noting that, in Newton's view, an *immaterial* agent—that is, an agent that comprises no matter or energy—could mediate gravitational interaction at a distance. (See Newton's statement quoted earlier in this section.) The physical meaning of a gravitational field will be discussed in section 19.

In a space comprising *true* vacuum nothing, not even the potential for gravitational interaction can exist. A grain of sand placed in a true vacuum would experience no gravitational interaction. (Nor would a charged particle experience any interaction if all charged objects in the universe were too far away for an electrostatic force to be generated. This scenario requires the assumption that the range of electrostatic interaction is finite, which would make it analogous to the finite range of gravity.) Thus, it is only in a true vacuum that an action-at-a-distance mode of interaction is impossible. True vacuum means a portion of space in which there exists no medium, *either material or immaterial*, that could mediate a physical interaction. If two mass objects are separated by a space that comprises a true vacuum, no gravitational interaction can occur between them. (Note that gravitational potential energy in true vacuum cannot exist either.) In sections 37 and 38, I will suggest that true-vacuum regions have to exist in the universe. They can exist because the range of gravity is finite according to the infinite-time hypothesis and because the nature is simple according to the FLH.

10 Simplicity-of-Execution Principle

The simplicity-of-execution principle, which is stated and discussed in this section, is the most outstanding, direct consequence of the FLH. It shows that nature, in executing her laws, is simple in the extreme. That simplicity will help to explain the effect of gravitational interactions, the difference in the magnitudes of electrostatic and gravity forces, and the constancy of the speed of light.

Let one end of an insulated metal rod be heated for a period of time such that the condition of thermal nonequilibrium arises (see fig. 1). A PERFECT SINK with respect to heat flow exists at the left end of the rod such that the temperature at that end (T_1) is constant. In accordance with the SLT (law I in section 7), the nonequilibrium must decrease after that time—that is, after the heat source has been removed (fig. 2). That means that a certain action must be enforced by nature to eradicate the thermal nonequilibrium. According to the FLH, that action must be independent of the dimensions; the chemical composition; the internal structure, or the mass of the rod; the degree of nonequilibrium; and the motion of the rod. Otherwise, nature would have to make measurements, and some principles would have to be built into the SLT to allow for decision making and for enforcing different actions depending on the rod's properties, dimensions, and/or motion. Such built-in principles would contradict the FLH. It follows that the SLT is executed—that is, the

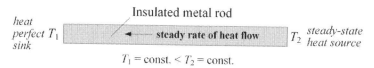

T_1 = const. < T_2 = const.

Fig. 1 Constant degree of nonequilibrium in a steady-state process. Owing to conflicts with heat source and heat sink (conflicts with steady heat energies at both ends of the rod), nonequilibrium is not eradicated.

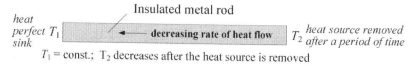

T_1 = const.; T_2 decreases after the heat source is removed

Fig. 2 Nonequilibrium is eradicated as required by law I after the heat source is removed. The quantitative effect does not depend on law I. It depends on the properties of the rod and temperatures T_1 and T_2.

eradication of nonequilibrium is enforced—in exactly the same way regardless of the rod's dimensions, its properties, or any other relating circumstances. That means that if another rod made of a different metal and having different dimensions was heated to a different degree of nonequilibrium, the action enforced by the SLT would be exactly the same as in the case of the first rod.

What transpires here is that, from the perspective of the quantitative response of matter or energy, an FLP is executed without any relation to physical settings, since nature has no means, according to the FLH, by which to calculate the amount of the influence of interfering matter or energy or the amount of the influence of interfering FLPs—that is, the influence of a given physical setting. Similarly, according to common sense, nature has *no interest* in calculating the amount of the influence of interfering matter, energy, or interfering FLPs. The SLT is executed precisely according to the statement of the law: "nature eradicates nonequilibrium between physical entities," and it is impossible to draw any quantitative conclusions in respect to that law or, specifically, to determine the amount of action enforced by it. The quantitative differences in the dynamics of heat flow in two different rods heated to different degrees of nonequilibrium are independent of the enforcement of the SLT itself. They result from different properties of the rods, different degrees of nonequilibrium, and/or other interfering FLPs, including, for example, the laws that govern the kinetic energies of atomic and molecular motions in a solid body.

As another example, consider law II, "mass resists a change to its state of motion," which is a nonquantitative statement, as it says nothing about the amount of resistance or the quantity of mass. Thus, it is not possible to conclude from law II that, for instance, resistance to change in the motion of a mass object increases in proportion to its mass when an unbalanced force of a certain magnitude acts upon it. The execution of law II in itself does not result in any quantitative effects. That law is always executed in the same way regardless of the mass of an object. Otherwise, nature would have to first measure the mass and then decide on the quantity of the inertia force (the amount of resistance), which would contradict the FLH. If an unbalanced force is exerted on an object, the object will no longer move with constant velocity but will instead accelerate with an amount of resistance dependent on its mass. The amount of acceleration will be determined by the amount of conflicts with other interfering FLPs and/or other matter or energy.

The foregoing examples can be generalized by the *simplicity-of-execution principle*, stated as: FLPs enforce nonquantitative actions—for instance, nonequilibrium between physical entities is eradicated, or two like charges repel each other—as stated in laws I–V; and forbid certain events—for instance—energy cannot be created—as stated in laws VI–X; and cause nothing else. Any observable quantitative phenomena—that is, the quantitative response of matter or energy to the execution of an FLP—are caused by other interfering FLPs and/or interfering matter or energy, which is to say that they are caused by *conflicts*. The simplicity-of-execution principle is a direct consequence of the FLH.

A pen laid on a desk is a simple example of a conflict between the execution of an FLP and interfering matter. The desk, which represents interfering matter, is in conflict with law III, which requires the pen to move down, toward the earth. The quantitative effect of that interference is represented by the force that the pen exerts on the table, which depends on the mass of the pen—that is, on the amount of conflict. A more complex example of such a conflict is illustrated in figure 1. While law I requires thermal nonequilibrium to be eradicated, the temperature T_1, which is constant at the left end of the rod (at the heat sink), and the heat source, which is at the right end of the rod and which maintains a constant temperature T_2, prevent eradication. In this example, the energy source and the sink interfere with the execution of law I, while the other matter (the metallic matter of the rod) interferes with the rate of the energy flow that results in a quantitative effect: a finite rate of energy flow. A more general example of a conflict with the execution of an FLP relates to the executions of laws III, IV, and V. The statements of those laws require instantaneous actions. For instance, the statement of law III requires that two objects move toward each other with infinite speeds. We know, however, that those speeds are finite. The finite speeds result, as a minimum, from the conflict with the simultaneous execution of (interfering) law II, which requires the two objects to resist motion.

In the case of laws III, IV, and V, the conflicts determine the amounts of acceleration experienced by two mass objects or two charges. So far, we do not know why those accelerations have the values that they have. For instance, in the case of law III, the acceleration of an object depends on the value of the gravitational constant (G) according to Newton's law of gravitation and to his second law of motion. The value of G is set by the conflicts with other interfering FLPs and interfering matter and/or energy. We did not determine the value of G from a theory. That value was derived from experiments and observations. (Possible interpretation of constant G will be discussed in section 22.) In summary, the FLP that causes gravitational interaction is very simple: two mass objects attract each other. Nature executes that law. The execution of that law is affected by the conflicts that result in observable quantitative effects. The quantitative effects of gravitation can be described by Newton's law of gravitation, which is not a fundamental law of nature (it is much too complex for nature to know it and then execute.) Similar conclusions apply to the execution of other FLPs.

Within the current paradigm of physics, the distinction between the physical events caused by the execution of an FLP and those caused by the interfering FLPs and/or interfering matter or energy does not seem to be recognized. I believe that this lack of recognition presents a prime impediment to our understanding of laws in physics. To appreciate the importance of that distinction, consider the formation of a star, which is of particular relevance to the theme of this book. The infinite-time hypothesis implies that at least some stars are formed out of radiation emitted by other stars, which was a conclusion drawn in section 3. That is possible only because of the SLT requires the mass of a star to be dispersed—and it is apparently dispersed in the form of thermal radiation as I will discuss in the next section—so that emitted radiation is available for the formation of new stars. On the other hand, while the SLT implies that a star should not be formed as the increase of nonequilibrium in the universe that is caused by the formation of a star should not happen, the SLT is in conflict with the gravitational and other forces that bind the matter. Those conflicting forces prevail, and a star is formed.

The simplicity-of-execution principle reveals that there are only two rules enforced by FLPs:

- Rule 1, which underlies the *action* laws I – V: nature eradicates, or resists, nonequilibrium (nonstability) or, in the case of law III, nature drives two masses toward a higher nonequilibrium state. For instance, according to law IV, nature forces two unlike charges to come together such that the two-particle system is driven toward a stable (equilibrium) state. The system will be in equilibrium when the particles collide and the system becomes neutral. Analogous explanation applies to law V.° In the case of law II, mass resists nonequilibrium caused by the application of an unbalanced force, and a balancing force (the force of inertia) is exerted as a result of the execution of law II. In this regard, note that a mass object in motion, with no "unbalanced" force acting upon it, is in equilibrium. It appears, therefore, that laws II, IV and V represent special cases of the SLT. If so, there would only be two action laws executed by nature at the macroscopic level: law I and Law III. While law I (the SLT) drives matter toward equilibrium, the opposite law III (the law of gravitation) drives matter toward the concentration of mass—that is, toward nonequilibrium. The opposite actions resulting from the executions of laws I and III will be discussed further in section 25.

- Rule 2, which underlies the *conservation* laws VI – X: fundamental physical quantities cannot be created or destroyed. (As pointed out previously, a more appropriate statement of the conservation laws is: "nature has no hidden sources of, or storage bins for, physical quantities.")

In summary, the FLH allows for distinguishing between the nonquantitative effects caused by the executions of FLPs and the quantitative effects resulting from conflicts with those executions. The nonquantitative executions of FLPs reveal nature (physics) at the fundamental level to be extraordinarily simple.

11 Second Law of Thermodynamics

In the preceding section, I used the SLT as an example of an FLP to conclude that the executions of FLPs are nonquantitative. Therefore, I

° With respect to the interaction of two like charges resulting from the execution of law V, I expect that the range of electrostatic interaction is finite, similar to the finite rate of gravity. This is for a finite charge causing an infinite effect would pose a paradox. Therefore, two like charges will be moving apart until the interaction between the charges ends, at which time they will become stable—that is, they will be in equilibrium with respect to each other.

have to confirm that the SLT is indeed an FLP. I am raising this issue because, as pointed out previously in section 7, the common belief is that the SLT is valid only statistically. Most likely, that belief stems from Boltzmann's statistical interpretation of that law, which normally is stated as: "There are so many more disordered states than ordered ones." Boltzmann's interpretation implies that, in (very) rare cases, an isolated, disordered system can transform itself into a more ordered state. If so, the SLT would not be exact and absolute, which also means that it could not hold the status of an FLP.

In support of the permissible violations of the SLT, a number of "thought experiments" have been proposed. The one that is perhaps the most popular involves a closed system (a box) comprising two equal-sized compartments and containing a large number (N) of gas molecules. For ease of discussion, assume N to be an even number. At the start of the experiment, a removable wall separates the compartments, and all molecules are contained in one of the compartments. The gas contained in the system (in the box) is in a state of strong nonequilibrium. The separating wall is then removed, and, in accord with the SLT, the system of molecules reaches equilibrium at a time t_1—that is, at the time when each compartment comes to contain $N/2$ molecules. However, at a later time (t_2), one may find the distribution of molecules at, say, $N/2 + 3$ in one compartment and $N/2 - 3$ in the other compartment. On the face of it, one might then conclude that the SLT was violated between times t_1 and t_2—that is, the order in the closed system was increased between those times. That conclusion, however, would be erroneous. This is for the execution of the SLT was affected by conflicts that (temporarily) prevailed over it. In accordance with the kinetic theory of gases, a gas molecule moves randomly with a molecular velocity dependent on the temperature of gas. Therefore, there was, on average, a conflict between the execution of the SLT and the kinetic behavior of the gas molecules that resulted in an illusory violation of the SLT. The random velocities of the gas molecules indicate that, during the t_1–t_2 time period, there were more molecules moving toward one compartment than toward the other. As a result, the SLT *appeared* to be violated, since the conflict between the execution of that law and the random motions of gas molecules was not accounted for. To make this explanation clearer, consider an iron filing falling toward the earth as a result of the execution of law III (the gravitational-interaction law). Let a strong magnet be placed just above the falling filling, with the magnet being stationary in the earth's frame of reference. The filling would be strongly attracted toward the magnet and would move upward toward the magnet—that is, the magnetic force acting on the filling would be stronger than the force of gravity. Does this mean that the law of gravitation has been violated? Certainly not. It merely means that the conflicting electromagnetic interaction law overcame the effect of law III. The same argument applies to violations of the SLT observed in conjunction with conducting any other thought experiments. Those violations appear to be illusory only. I believe that if all conflicts with the executions of the SLT are accounted for, no violations of the SLT would be found.

The above argument also applies to actual experiments. The authors of a 2002 paper claimed that the SLT was violated in an experiment that involved moving a micron-sized latex particle in water.[90] It was determined that, at times, the energy of the particle increased as a result of its interactions with water molecules (i.e., the disorder of the system decreased), so the SLT was violated. Again, the SLT was not really violated, since the increase in the energy of the latex particle was caused by a conflict with interfering matter and energy—the motions of the water molecules.

Not all physicists agree that the SLT is valid only statistically. For instance, Alfred B. Pippard states:[91] "Although very few hypothetical experiments...have been analyzed in...detail [those experiments were designed to show violations of the SLT], it appears most probable that they all fail to violate the SLT on account of the necessary entropy generation by the observer who controls the process. There is thus no justification for the view, often glibly repeated, that the SLT is only statistically true, in the sense that microscopic violations repeatedly occur but never violations of any serious magnitude. On the contrary, no evidence has ever been presented that the SLT breaks down under any circumstances." In that statement, Pippard refers to the unaccounted-for influence of an observer on the result of a hypothetical experiment that falsely implies a violation of the SLT. I want to emphasize that I am in a complete agreement with Pippard.

In developing the GPU theory, it will be essential to recognize that the SLT is indeed an FLP. As such, it is also exact and absolute. That means that it applies to the nonequilibrium of *any* physical quantity. It will be particularly important to recognize that it applies to the nonequilibrium of mass. To appreciate that notion, hang a massive (e.g., iron) ball at the center of a room. You will see a highly ordered state of the distribution of mass. A lot of mass is contained in the ball, and relatively little mass is contained inside the remainder of the room (in the ball's surroundings). Since the SLT drives an ordered system into a more disordered state, the mass of the ball should continuously be dispersed into the surroundings. And it does. Recall that any object with a temperature above absolute zero emits thermal radiation (energy). And you must have heard that according to the theory of special relativity, mass and energy are one and the same. Therefore, it appears that the SLT drives the mass in the room toward a more disordered state by means of thermal radiation emitted by the ball.

If the SLT is indeed absolute, it could be seen responsible for the observed, large-scale homogeneity and isotropy of the distribution of matter in the universe. This will be discussed further in section 25.

12 Other Laws in Physics

The discussion presented in section 10 suggests that a key to the understanding of nature (physics), which includes the understanding of the universe, is the appreciation of the fact that there are two kinds of laws in physics: fundamental laws of physics and nonfundamental laws, which will be referred to henceforth as "physical laws." The former are exact and absolute; have no underlying principles; do not incorporate the concepts of space and/or time; and exist independent of the human mind. The latter are stated by mathematical formulae that describe quantitative relations between the properties of matter, energy, space, and time. They explicitly account for the quantitative effects of conflicts between the execution of an FLP and the execution of interfering FLPs or interfering matter/energy. Physical laws are invented by physicists and exist in the human mind only. (Many scientists believe that to be true. For instance, Davies is of the opinion that "the statements called [physical] laws .... clearly are human inventions."[92]) I want to emphasize that it is the statements of physical laws, and not the physical phenomena those laws describe, that exist in the human mind only. The important point made in section 10 was that the observed quantitative phenomena were not the result of the execution of an FLP. They result from conflicts between the execution of an FLP and other interfering FLPs and/or interfering matter or energy. *Physical laws are not executed.* They describe, and can used to predict physical phenomena. That statement, it seems, presents a rather uncommon look at physics, the importance of which cannot be overstated. Another important point is that it is the physicist, and not nature, that has an interest in formulating a physical law. "Dehumanized nature" means, as I said earlier, that nature that has neither an interest in describing or predicting physical phenomena nor a curiosity about how the universe works.

An examination of the FLPs identified in section 7 also reveals that it is not possible to ascertain that a physical law—that is, a quantitative relation between physical quantities that incorporates the concept of space or time or both—is exact. That is because the FLPs, which underlie physical laws, are nonquantitative and do not incorporate the concepts of space or time. Therefore, it is not possible to use an FLP to prove that a physical law is exact. Any such law needs to be considered approximate, even if the results of experiments suggest that it is perfectly accurate. This supposition is not new. Henri Poincaré, and other scientists, thought long ago that laws in physics must be approximate (I think they referred to "physical laws" as opposite to "fundamental laws of nature.)" On this subject, Poincaré and Halsted wrote in 1907, "If we look at any particular law, we may be certain in advance that it can only be approximate."[93] More recently, Feynman wondered if the approximate character of physical laws represents a property of nature.[94] The above remarks on physical laws suggest that their approximate character is not a property of nature but rather a property of the statements of such laws.

In contrast, there are no approximations allowed for in FLPs, which are meant to be exact. In this regard, Norman Swartz[95] states that "if scientific laws are inaccurate, then—presumably—there must be some other laws ..., doubtless more complex, which are accurate, which are not approximation to the truth but are literally true." (In the terminology used here, "literally true" corresponds to "absolute.") Swartz's statement contradicts the FLH. For instance, the theory of general relativity is more complex than Newton's law of gravitation, yet, as pointed out above, it is impossible to assert that it is exact. General relativity and all other quantitative theories of gravitation would have underlying mechanisms, principles, and/or other laws and as such would have to be categorized as physical laws. Accordingly, it is not possible to make an assertion about the exactness of any of those theories.

The search for a theory of everything (a master law of physics) has become the proverbial holy grail of today's physics, and many physicists (e.g., Smolin[96]) consider it a necessary goal. However, the existence of a fundamental master law, from which all other FLPs could be derived, would contravene the FLH, for such a law would have to incorporate some principles and decision-making abilities. Then, the master law in itself could not hold the status of an FLP, whence no fundamental law of physics would exist. Consider, for instance, the electrostatic interaction laws IV and V. If a master law existed, the execution of it would have to incorporate a measurement skill that would allow for identifying if the two charges are like or unlike and then deciding which of the two laws were to be applied. That would contravene the FLH. Therefore, one concludes that laws IV and V are independent of each other. The highly accurate mathematical expressions of the corresponding physical laws—both shown by equation (18)—suggest that the interfering FLPs and the interfering matter/energy affect the interaction of the like and the unlike charges in quantitatively the same, while opposite ways. However, that does not mean that there is one fundamental law underlying electrostatic interactions. (Nevertheless, if laws IV and V are recognized as different aspects of the SLT—as suggested in section 10—then, the SLT becomes indeed the master law underlying electrostatic interactions.)

Furthermore, it appears that a master law would have to be limited to a class of physical phenomena only—for instance, to the forces of nature. This is for there seems to be no evidence of any kind that law I (the SLT), law III (the law of gravitational interaction), and law VI (the momentum-conservation law) could be derived from a single master law. That does not mean that a *master physical law* cannot exist. It may be found one day in the form of a mathematical model (of great complexity) from which all other physical laws could be derived.

In this section, nature (physics) was further dehumanized, consistent with the implications of the FLH. Specifically, physical laws, which are expressed by mathematical formulae that describe, and allow to predict relations between physical quantities, space and/or time, were asserted to exist in the human mind only.

(There are other aspects of physics that have been humanized by scientists. Of particular interest to this book is Olbers's paradox, which states that in a spatially-infinite universe, the entire night sky should be as bright as the surface of the sun. For Olbers's paradox to be rational, stars/galaxies that emit radiation at frequencies other than the visible-light frequency cannot exist. Put otherwise, the assumption that underlies Olbers's paradox is that all stars/galaxies emit radiation at the visible-light frequency. Without any doubt, that assumption was made because the human eye can detect visible radiation only. (At the time of Olbers, scientists were not aware of microwave or X-ray radiation.) Thus, the belief that the night sky should be as bright as the surface of the sun presents a striking example of the humanization of nature. If invisible-to-human-eye stars/galaxies that emit radiation at frequencies other than the visible-light frequency do actually exist, we would expect to see them at night as black elements of the sky. The night sky in a spatially-infinite universe would be expected to look the same as it actually does.)

13 Space and Time

The fact that none of the fundamental laws of physics identified in section 7 incorporate the concepts of space (distance) or time presents a somewhat unorthodox view of physics. An explanation of those concepts with reference to the two kinds of laws in physics (discussed in the preceding section) will therefore be useful. In this section, I will discuss the concepts of physical time and physical space, which are incorporated in physical laws yet absent from the fundamental laws of physics. In that discussion, I will refer to two most astonishing traits of the human mind: (-) the ability to correlate many unrelated physical times and interrelate them by constructing a single human time, and (--) the ability to identify many unrelated physical spaces and relate those by constructing a single human space. I want to emphasize that it is not my intention to contribute in any way to the understanding of what time or space is. What time and space are has been debated by some most brilliant philosophers and physicists for more than two thousand years and still, clearly, that debate has not been settled in any way. The following discussion of the concepts of space and time is limited to the explanation of why the time and space variables appear in the statements of physical laws while the statements of the FLPs do not account for those variables.

As FLPs are independent of time, laws I, III, IV and V imply that the motions of matter and energy enforced by the executions of those laws should be instantaneous (time is irrelevant), which is not possible because of conflicts with other interfering FLPs and/or interfering matter or energy. That suggests that a physical time comes into existence as a result of such conflicts. For instance, law III requires two mass objects to move toward each other instantaneously. That is not possible, because law II requires those objects to resist a change in motion, whence physical time comes into existence as a result of that and other conflicts. Physical time will cease to exist when the two objects collide. Another example is a metal rod in a state of thermal nonequilibrium caused by a heat source placed adjacent to one end of the rod. The SLT requires the nonequilibrium to be eradicated instantaneously. That is not possible, because the motion and energy-conservation laws and other conflicting laws are executed inside the rod, and so the nonequilibrium resists instant eradication. Again, physical time will cease to exist when equilibrium is reached. Aristotle thought that time exists only when there is a change. (According to Aristotle, "But neither does time exist without change; for when the state of our own minds does not change at all, or we have not noticed its changing, we do not realize that time has elapsed."[97]) I suggest the same idea here in regard to physical time, and clarify that a "change" means a physical interaction resulting from the execution of an FLP. It follows that there are an infinite number of independent physical times (in a spatially-infinite, homogeneous universe).

From the perspective of physical laws, the concept of space is associated with, and similar to the concept of time. For two objects attracted by an interaction, physical space exists for as long as the objects are separated. When the two objects collide and thus form one object, physical space ceases to exist. In the case of two like charges, space (and time) will cease to exist when the two charges become separate enough for the electrostatic interaction to cease. When the equilibrium of a physical quantity between two locations is reached (law I), physical space ceases to exist as well. Just as there are an infinite number of physical times, there are an infinite number of independent physical spaces (in a spatially-infinite, homogeneous universe).

Consider an apple falling from a tree and a metal rod in thermal nonequilibrium, with the heat source removed. The time it takes the apple to fall to the ground and the time it takes the rod to reach equilibrium are entirely independent of each other. The human mind can correlate those two events by using a human clock, which defines human time. A similar implication applies to space.

If physical time and/or physical space were incorporated in a law of nature, she would have to measure time and/or space to execute that law, whence such a law could not be taken to be fundamental in the sense of the FLH. While executing an FLP, nature does not know that physical space and physical time are in existence. Since she has no memory, she does not know that in the previous instant, the FLP was not executed instantaneously (whence space and time came into existence). In the next instant, nature will enforce the FLP in exactly the same way, not realizing that in the prior instant, two mass objects did not collide instantaneously, and nonequilibrium was not instantaneously eradicated. (The phrase "next instant" refers to human time. Nature, having no memory, doesn't know the next from the present or the previous instant.) Thus, at the fundamental level of physics, the concepts of physical time and physical space do not exist. This emphasizes that classical physics is extraordinarily simple at the fundamental level.

To further appreciate the concepts of physical as opposed to human time and space, consider the following scenario: Your friend drops a stone from a balcony. Law III requires that the stone and the earth attract each other, so the stone and the earth move toward each other. The stone is seen "falling." Prior to the start of the fall, the stone and the earth were attracting each other, but your friend, by keeping the stone in her hand, caused a conflict that prevented the execution of that law. At the moment when your friend let the stone go, physical time and physical space came into existence. You are under the balcony and catch the stone. Physical time and physical space cease to exist due to the new conflict, which is your hand catching and holding the stone. Yet human time and human space existed in your mind before the stone was dropped, throughout its fall, and after you caught it, since you were able to observe both the motion of a human clock's hand and the distance between the stone and the surface of the earth.

In conclusion, physical space and physical time have meanings only when physical interactions are occurring, which appears to correspond well to Aristotle's original conception of space and time. This view of space and time would also correspond to Leibniz's "systems of relations between things"[98] if those "relations" were interpreted as "physical interactions," which is the relationist's position. The idea of space that exists only when two objects interact is often attributed to George Berkeley, who expressed his thoughts, "So that to conceive Motion, there must be at least conceived two Bodies, whereof the Distance [space] or Position in regard to each other is varied. Hence if there was one only Body in being, it could not possibly be moved. This seems evident, in that the Idea I have of Motion doth necessarily include Relation."[99] Herein, I consider Berkley's concept of "motion" to be equivalent to Aristotle's concept of "change." It follows that, with regard to physical laws, Newton's proposition of one absolute time and one absolute space that exist in a physical sense, which is a substantivalist's position, cannot hold if the FLH holds.

I believe that one of the difficulties with comprehending the nature of space and time arises because there are infinite numbers of independent physical spaces and independent physical times, but in our minds, we assume that only one space and one time exist. In developing special relativity, Einstein showed that each observer carries his own time and his own space.[100] He therefore implicitly implied the existence of infinite numbers of times and spaces (in a spatially-infinite universe). Herein, I look at the infinite numbers of times and spaces from a similar perspective, which indicates that infinite numbers of physical times and physical spaces are the results of infinite number of physical interactions that are independent of one another. The existence of independent physical times is trivial: the time it takes an apple to fall from a tree is entirely independent of the time that one needs to drive to work. By the same argument, the existence of independent physical spaces also is trivial: each set of two interacting objects has its own space (distance) that, in general, is independent of the distances between other interacting objects. One can appreciate that argument by considering two sets of interacting charged particles (charges). Each set comprises two charges. One set is located in the Milky Way galaxy and the other set is located in the Andromeda galaxy. The distances between the two interacting charges in each set of are completely independent of each other.

In any physical law that describes a specific phenomenon, space and time have to be accounted for always in the same way, independent of the location in human space or human time. This means that physical laws have to be the same throughout the human time and throughout the universe. Put otherwise, the same conflicts that affect the execution of an FLP are expected to cause the same response of matter and energy regardless of the location in human space or human time. The underlying suppositions are that the FLPs are the same regardless of the location in human space and time, and that the universe is homogeneous on a large scale. The latter supposition is a consequence of the perfect cosmological principle. It follows, that human space and human time play no role in physical phenomena. They merely serve to describe and predict physical phenomena in a way that is intelligible to the human mind.

In summary, the common concepts of space and time appear to have no meaning at the fundamental level of physics, while those concepts are indispensable to formulating the quantitative descriptions of physical phenomena—that is, to the physical laws. As pointed out previously, the abilities to correlate the infinite number of physical times to create one single human time and to correlate the infinite number of physical spaces to create one single human space are two of the most amazing abilities of the human mind. The shortcoming of nature is that she does not have such abilities. Human time and human space exist in the human mind only. Nevertheless, since they can be directly measured, they can be considered to be physical entities when formulating a physical law.

Because they are inherent to physical interactions, physical space and time have to be taken to represent physical realities for the purpose of formulating physical laws. This is because, similar to human time and space, physical space and time can be measured. And, if something can be measured by all observers with exactly the same result (accounting for relativistic effects, of course), that something can be taken to represent a physical reality regardless of the fundamental laws of nature that do not account for the concepts of physical space or time.

14 Physics and Mathematics

Perhaps the most controversial challenge to the current paradigm of physics is the dehumanization of nature with respect to her ability to know and apply mathematics. While some physicists and philosophers of science would agree that nature does not know mathematics, it seems that the majority of physicists would argue otherwise. In their view, widely-accepted physical laws ("mathematical models") such as the theory of general relativity or the standard model of particle physics incorporate an implicit inference that mathematics is somehow built into the fabric of nature. In other words, "the laws of nature must have been formulated in the language of mathematics.[101]" As pointed out previously in section 6, if general relativity in the form of Einstein's field equations is considered to be a law of nature that exists independent of the human mind, then nature would have to solve an unimaginably large number of sets of ten simultaneous, nonlinear partial differential equations in each infinitesimally small period of time, which would support the notion that mathematics is built into the fabric of nature. If, however, general relativity merely represents a mathematical description of gravitational interactions that exists in the human mind only, then there is no need to grant nature the powers of an almighty mathematician. Thus the relevant question is: is the idea that mathematics is built into the fabric of nature a rational notion? An answer to this question is crucial, since mathematics being or not being built into the fabric of nature leads to two utterly different understandings of the universe.

The idea that mathematics is intrinsic to nature's laws has been suggested since ancient times, first, perhaps, by the early Greek philosophers and certainly by Plato and Pythagoreans.[102] In recent times, Galileo was likely the first to advocate an intrinsic relation between mathematics and physical laws. The view of modern mathematicians on the role of mathematics in advancing physics is well summarized by the comments made by Morris Kline[103] on the paper by Lewis Pyenson:[104]

> A key point of the paper is the difference in approach to physical problems taken by mathematical physicists as opposed to theoretical physicists. In a paper published in 1908 Minkowski reformulated Einstein's 1905 paper by introducing the four-dimensional (space-time) non-Euclidean geometry, a step which Einstein did not think much of at the time. But more important is the attitude or philosophy that Minkowski, Hilbert,...Felix Klein and Hermann Weyl pursued, namely, that purely mathematical considerations, including harmony and elegance of ideas, should dominate in embracing new physical facts. Mathematics so to speak was to be master and physical theory could be made to bow to the master. Put

otherwise, theoretical physics was a subdomain of mathematical physics, which in turn was a subdiscipline of pure mathematics. In this view Minkowski followed Poincaré whose philosophical view was that mathematical physics, as opposed to theoretical physics, can furnish new physical principles. This philosophy would seem to be a carry-over…from the Eighteenth Century view that the world is designed mathematically.

The idea of advancing physics by the use of mathematics reflects the highly influential view of Nobel Laureate Paul Dirac, who wrote that "it is more important to have beauty in one's equations than to have them fit experiment."[105] One of the most influential publications on the role of mathematics in physics is an essay written by Eugene Wigner.[106] While Wigner's view was that the laws of nature may have been formulated in the language of mathematics (which was Galileo's idea), he also thought that idea to be a mystery. The view that mathematics is built into the fabric of nature has been shared by some of the most accomplished physicists (e.g., Penrose[107], Wilczek[108].) Yet despite the numerous discussions by physicists of the role of mathematics in physics that were held over the last fifty years, no consensus on that role has been reached, as discussed by Davies.[109]

The problem with Wigner's essay is embedded in the very title of the essay, "The Unreasonable Effectiveness of Mathematics in the Natural Sciences." (That title is essentially a rephrased question of Einstein who asked in 1921: "How can it be that mathematics, being after all a product of human thought which is independent of experience, is so admirably appropriate to the objects of reality?"[110]) In order to imply that something is unreasonable, an alternate "reasonable" approach must exist. Yet no other—that is, nonmathematical—approach to advancing physics has been found. The only known way to advance applied physics is to use mathematics—that is, to develop mathematical models designed to describe physical phenomena. (Mark Steiner expresses a similar opinion: "Scientists used mathematical analogies because they had no real alternative."[111]) And since this is the only known approach, the question of whether the effectiveness of mathematics is reasonable has no meaning. While the indispensability of mathematics to advancing applied physics cannot be denied, this does not necessarily mean that mathematics is built into the fabric of nature.

Wigner's notion reflects the platonists' ontological interpretation of the Quine-Putnam indispensability argument (put forward by Willard van Quine[112] and Hilary Putnam[113]). Similar to some physicists, platonists argue that since mathematics is indispensable to advancing physics, we must recognize mathematical entities as physical realities, since we ought to believe in our best theories of science, which is a statement of the Quine-Putnam indispensability argument (see Mark Colyvan[114]). Some philosophers of science agree with and defend the indispensability argument with further arguments (e.g., Alan Baker,[115] John Burgess,[116] Bob Hale and Crispin Wright,[117] Mark Colyvan[118]), while others (nominalists) argue otherwise (e.g., Mark Steiner,[119] Hartry Field,[120] Elliott Sober[121]). Like the physicists, the philosophers have not reached any consensus in this regard.

As discussed in section 12, the platonists' "best theories of science" are, in fact, the mathematical models (physical laws) formulated by physicists, which exist only in the human mind only. They do not exist as physical realities. This was concluded based on the premise that the FLH holds true. Thus, the existence of mathematical entities as physical realities, which is the Quine-Putnam indispensability argument, is not pertinent. (Note that the FLPs are not relevant to the present discussion as they do not contain mathematical entities.) This suggests that the indispensability argument presents not a real argument but a belief. That is the same problem that physicists face.

The FLH bluntly rids mathematics of any role in physical phenomena. That is admissible, as the question of whether mathematics is built into the fabric of nature appears to be a question of belief only. Aesthetics (simplicity) aside, it follows that the only judge here would be the efficacy of that belief. I want to state again that the FLH does not deny the indispensability of mathematics to advancing physics. It implies, however, that no mathematical entities exist independent of the human mind.

Table 1 summarizes the relation between physics and mathematics presented from the perspective of formulating physical laws. It illustrates similar structures of physics and mathematics, which, I believe, explains why mathematics is so effective in formulating physical laws.

Table 1: Physics versus mathematics.

Fundamental Laws of Physics	Fundamental Axioms of Mathematics
Independent of one another (there is no master law)	Independent of one another (there is no master axiom)
Physics	**Mathematics**
Fundamental laws of nature underlie physical laws (physical relations)	Fundamental axioms of mathematics underlie mathematical relations
Physical relations are relations between physical quantities (represented by numbers)	Mathematical relations are relations between mathematical quantities (numbers)
Physical relations are invariant in space and time; they hold the same under the same physical settings and are objective (independent of the observer)	Mathematical descriptions can be made invariant in space and time; they hold the same under the same mathematical settings and are objective (independent of the observer)
Dozens of physical laws (mathematical relations) are required to describe nature	An infinite number of mathematical relations (equations, etc.) are or can be derived and made available to describe nature
Execution by Nature	**Execution by Mathematician**
Fundamental laws of nature are executed independent of the human mind	Mathematical relations are executed (derived) independent of nature
There are no choices in executing the fundamental laws of nature	There is a freedom to choose the best suited axiom, function, constant, etc.

The statements in the left and right columns of table 1 are made independent of mathematics and physics, respectively, and the indicated similarities as well as the convenient differences, which are identified in rows 5 and 7 in the table, are not related in any tangible way. The table suggests that mathematics is well suited for making quantitative predictions in physics, which is further argued for in the next paragraph.

Since physical phenomena are normally continuous in the sense that the magnitude of a physical quantity changes continuously with space, time or another physical quantity, physical laws expressed in the form of continuous mathematical functions are well suited for making predictions in physics. In the simplest case, interpolation or extrapolation (curve fitting) can be used to predict a magnitude of physical quantity. An example of a more complex prediction made using a mathematical relation is the discovery of the planet Neptune. The existence of Neptune was predicted using Newton's gravity force equation (see equation (9) in section 22), which presents a mathematical relation that describes gravitational interactions between the planets and the sun with a very high degree of accuracy. That relation, the use of which identified certain irregularities in the orbit of Uranus, allowed for predicting the existence of Neptune, which was the cause of those irregularities. However, the success of that prediction in no way implies that mathematics is built into the fabric of nature. It merely demonstrates that Newton's gravity-force equation (a physical law) is highly suitable for the description of gravitational interactions between the sun and the planets.

Using mathematical fields to describe physical interactions has been particularly successful. That is because such fields simplify the presentations of physical laws. For instance, the enforcement of law III between each pair of mass objects may result in multiline gravitational interactions. That leads to a rather complex presentation of a physical law of gravitation for more than two mass objects. To appreciate the success of the field concept, consider ten mass objects gravitationally interacting with one another, as illustrated in figure 3. The force acting on each object has, in reality, nine components. The dashed-line arrow shows the resultant of the forces acting on object 7, and the solid-line arrows illustrate its components. The ten objects can be assumed to generate a gravitational vector field around them. Each object is then assumed to generate a force vector at each point of space. The total field value at a point in space equals the sum of those vectors, which is a single vector (a mathematical gadget) that represents the force of gravity. That force would act on a new object inserted into space at that point. (The force of gravity acting on an object at a point in space is calculated by multiplying the mass of the object by the field value at that point.) Thus the presentation of the law of gravitation is simplified when using

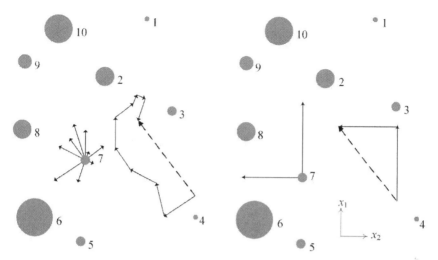

Fig. 3 Gravitational "action-at-a-distance" mode of interactions.

Fig. 4 Gravitational "vector field" mode of interactions.

the gravitational field concept, as can be seen from the comparison of figures 3 and 4.[p] The law of gravitation expressed in terms of a field is called Gauss law for gravity. It is fully equivalent to Newton's law of gravitation but is simpler when applied to multi-object situations. In some cases, the calculations performed using the Gauss law are simpler than those performed using the Newton law.

Physical reality of a field that comprises numbers would be in disagreement with the FLH: nature, in enforcing her laws, is unable to account for field values (numbers) that may vary with time and location. As I indicated previously, nature has neither a need nor a skill to use a gravitational field to calculate the force of gravity. The FLH does not imply, however, that a field in physics cannot exist as a physical reality. In section 19 I will introduce the concept of a gravitational field that, I believe, is physically real. Yet, that field will be nonmathematical—that is, there will be no field values (numbers) associated with that field.

---

[p] This discussion refers to the very difference between the views of platonists and nominalists. Platonists believe that a force vector (that I have called a "mathematical gadget") exists at each point of space regardless whether there is a mass object at that point or not, that is, they believe the vector to exist as a physical reality even if, at that point, the space is empty of mass and energy.

Regardless, I believe that conscious pretending that mathematics is built into the fabric of nature is the most appropriate approach to advancing applied physics. As the experience shows, such an approach is highly effective, and no other approach to advancing applied physics has been found. However, explicit recognition of the difference between nature and mathematics would be advantageous. For instance, one could not say that a physical law were correct or incorrect if one recognized that such a law represented, in fact, a mathematical description rather than a physical reality.[q] It is only the accuracy and the range of the applicability of that law as well as its consistency with other laws in physics that could be used to judge if that law was "correct" or not.

15 Physics in Trouble

Some physicists (e.g., Woit[122]) think that today's physics is in trouble, but no one seems to present a reason for that trouble at a fundamental level. One of the most comprehensive publications on the troubles with today's physics is the book by Lee Smolin[123], who, in 2006, pointed to the lack of progress in physics since the early 1980s. In particular, he stated, "But when it comes to extending our knowledge of the laws of nature, we have made no real headway." It seems that Smolin's statement was made in reference to the knowledge of physical laws, which do not exist as a physical reality, as I discussed in section 12. Smolin identifies five great troubles (problems) with today's physics. The FLH, however, suggests that those problems could be irresolvable in physical terms. They relate to the merging of physical laws or the explanations of some specific aspects of physics. For instance, one of the troubles that Smolin highlights is the lack of resolutions to problems in quantum physics. According to the FLH, however, the quantum theory presents a set of mathematical descriptions (a mathematical framework), and physically meaningful explanations of its aspects may or may not

---

[q] Contrary to most modern physicists, I think of mathematical physics to be *applied physics* rather than *theoretical physics*. In my view, theoretical physics is a scientific discipline concerned with the understanding of nature, its laws and the executions of those laws. Applied physics, on the other hand, is a scientific discipline concerned with physical laws that are necessary to describe and predict, in quantitative terms, physical phenomena. I consider theoretical physics to be the classical scientific discipline that used to be called *natural philosophy*.

exist. Thus, it might not be possible to comply with Smolin's appeal to "resolve the problems...by making sense of the [quantum] theory." (It is worth noting that an explanation of Newton's law of gravitation in physical terms has been sought for some 350 years, with no success.) Another one of Smolin's problems involves the question of how to merge the theory of general relativity (a deterministic description of gravitational interactions) with the quantum theory (a probabilistic description of subatomic interactions). According to the FLH, those theories represent mathematical models, and the widely sought-after merger would merely mean a merger of two mathematical frameworks adequate to account for both gravitational and quantum interactions. Similar comments apply to other problems with physics identified by Smolin. I suggest here that the trouble with physics is more fundamental than Smolin's problems: the trouble with physics is the humanization of nature by physicists whose foremost belief is that nature is the ultimate mathematician.

In case you are wondering, at this time, what this chapter II has been all about, here is a summary of it: it has been about nature, who just executes her laws (FLPs). She does nothing else. Conflicts associated with those executions arise and result in the observed quantitative behavior of matter and energy. Nature has no idea about those conflicts. She does not have any interest in describing or predicting physical phenomena; nor does she have any skills or means (scales, calculators, clocks, measuring sticks, etc.) to do so. Such interests and means are attributes of the physicist's mind. What's the big deal about all of this? Keep reading, and you'll find out how this will help you understand gravitational interactions, the equivalence principle, and the speed of light, all of which have to become the key components of any rational cosmological model.

CHAPTER III: GRAVITATION

To understand the cosmology of the infinite-in-time universe, it will be sufficient to read sections 16, 17, 19 and 20 of this chapter. To appreciate the issue of gravitational interactions in the infinite-in-time universe, the entire chapter should be read.

16 Background

The concept of gravitation incorporated in the GPU theory is explained in this chapter. As discussed previously in section 3, both the infinite-time hypothesis (in conjunction with Einstein's proof) and the only possible resolution of the gravity paradox imply a finite range of gravity. This implication is in disagreement with the oldest dogma of modern physics, the infinite range of gravity, which was first assumed by Newton in his law of gravitation in 1687.[124] Newton's assumption presents an extraordinary and probably unprecedented extrapolation of physical observations, as I will discuss in section 27. Furthermore, since Newton's law of gravitation is an empirical law, the infinite range of gravity can only be confirmed based on the results of astronomic observations, which is not possible, even in theory.[r] The infinite range of gravity is still taken for granted today, over three hundred years after Newton's *Principia* was published, and has been incorporated in general relativity and nearly all other theories of gravitation proposed over the years. In this chapter, the GPU concept of gravitation is explained, which is based on a finite range of gravity. It will involve the consideration of the infinite-time hypothesis and the simplicity-of-execution principle, which is a direct consequence of the FLH.

Owing to the above-mentioned disagreement, the discussion of gravitation presented in this chapter is rather intensive. It is designed to examine the implication of a finite range of gravity from various perspectives. The discussion is intended to show that a finite range of gravity, which is to become a key feature of the GPU cosmology, is not only logical but, also, in no violation of any results of observations or experiments. The conception of a finite range of gravity is not entirely

---

[r] While confirmation of a *finite* range of gravity is possible in theory, in practical terms, it is impossible as well.

new. It was proposed on a few occasions (e.g., by Allen Allen,[125] Peter Freund, Amar Maheshwari, and Edmond Shonberg;[126] Stanislaw Babak and Leonid Grishchuk[127]). Those proposals, however, were limited to the introductions of some mathematical modifications to the equations of general relativity and other theories of gravity. Herein, a finite range of gravity is implied on its own, independent of any law of gravitation.

In the ensuing discussion, wherever the word "mass" is used, it is intended to mean the rest (intrinsic) mass. In relativistic terms, the intrinsic mass of an object equals its total intrinsic energy divided by the speed of light squared, in accordance with the mass-energy equivalence relation, $m_{intrinsic} = E_{intrinsic}/c^2$. It will be important to keep in mind that the intrinsic energy of an object excludes the kinetic energy of its motion as a whole, which is referred to as the "translational kinetic energy," and any other energy that may depend on the location of the object (e.g., GRAVITATIONAL POTENTIAL ENERGY). That means that the intrinsic mass of an object does not depend on the motion of an observer or on the distance between the object and another mass object. However, intrinsic energy includes all other energies stored in a mass object, such as the kinetic energies of atomic electrons and molecules, as well as the nuclear, chemical and GRAVITATIONAL BINDING energies.

17 Finite Range of Gravity

In this section, frequent references are made to figures 5 through 8. I believe those figures to be simple. Yet one needs to study each of them carefully to make certain that everything illustrated in that figure is understood. Otherwise, the concept of a finite range of gravity, which is new, could be difficult to grasp. It will be of particular importance to understand why a finite range of gravity cannot be defined by a distance. It has to be defined by a volume.

A gravitational object, which I call a "g-object," is defined as an array of gravitationally interacting single-mass objects, which do not directly interact with any other single-mass objects. (The concept of "direct" versus "indirect" interaction will be explained shortly.) As an example, the gravitational system of a star, which could comprise the star and several planets, would form a g-object if it were located sufficiently far from all other stars such that no direct gravitational interactions would occur between any constituents of the g-object and other stars.

(The solar system would be a g-object if no other star directly interacts with the sun or the planets.) Because the range of gravity in the infinite-in-time universe is finite (section 3), the state of no direct gravitational interaction between the constituents of a g-object and an outside mass object can exist. Simply, if the distance between any constituent of the g-object and an outside mass object is larger than the sum of their gravity ranges, a direct gravitational interaction cannot occur between them.

Because of a finite range of gravity, there must exist a continuous boundary surrounding a g-object that delineates the range of gravitational interactions of the g-object in all directions. The space inside that boundary I will call the "gravitational field" of a g-object. Its volume I will denote v. The gravitational field of a single-mass object is illustrated in figure 5. A g-object comprising four single-mass objects, which I will call g-object m_{1234}, is illustrated in figure 6. The dashed circles, which illustrate two-dimensional sections of spherical gravitational fields, depict the boundaries of the individual fields of the four single-mass objects. The continuous line shown in figure 6 depicts the boundary of the gravitational field of g-object m_{1234} as a whole. In accord with the definition of a g-object, the gravitational field of g-object m_{1234} is not connected to any gravitational field of a single-mass object.

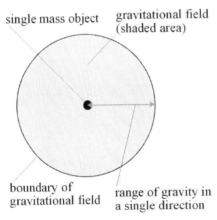

Fig. 5 Gravitational field of a single mass object.

The single object shown in figure 5 presents the simplest g-object. It comprises one mass object. In the ensuing, I will call an array comprising more than one mass objects a "g-object." I will call a single-mass g-object either a "mass object" or simply an "object."

As the magnitude of the gravitational field of a g-object equals the extent of its range of gravity, a direct gravitational interaction between the g-object and any single-mass object can only occur when their gravitational fields are connected. A new g-object is then formed as the constituents of the two original g-objects are

Gravitation

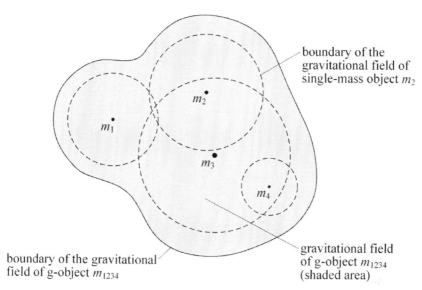

Fig. 6 G-object m_{1234} comprises four single-mass objects. Combined gravitational field of the four objects, outlined by the continuous boundary line, illustrates the gravitational field of g-object m_{1234}.

incorporated into a single g-object. This is illustrated in figures 7 and 8, in which the gravitational field of object m_5 gets connected to that of g-object m_{1234} (fig. 7) and a new g-object m_{12345} is formed (fig. 8). Note that the single-mass object m_5 could be replaced with a g-object comprising more than one single-mass object. Object m_5 does not *directly* interact with any of the four other objects since its gravitational field is not connected to any of the gravitational fields of those objects. It interacts with the other four objects *indirectly* only, via the fictitious object located at the gravity center of the five objects. (Direct versus indirect gravitational interaction will be explained further in section 20).

Examine the single-mass object shown in figure 5 from the perspective of a finite range of gravity. The supposition that the range of gravity of an object increases with an increase in its mass appears to be logical. That is because it would be rather untenable to assume that the range of gravity of an electron, which has a mass of 9.11×10^{-31} kg, is the same as the range of gravity of the sun, which has a mass of $1.99 \times 10^{+30}$ kg. From this consideration, the following postulate is put forward:

$$v = \frac{1}{D}m. \tag{1}$$

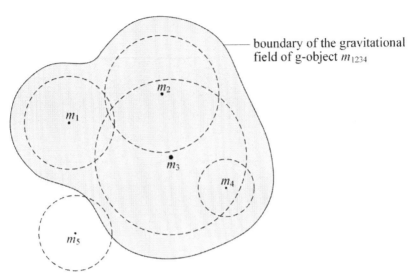

Fig. 7 Single-mass object m_5 interacts with g-object m_{1234}; the combined gravitational field is shown in figure 8.

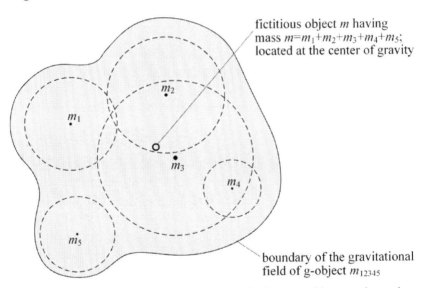

Fig. 8 Former g-object m_{1234} incorporates single mass object m_5 shown in figure 7 and becomes g-object m_{12345}.

In equation (1), quantity v represents the finite range of gravity. Its value equals the volume of the object's gravitational field. Quantity m represents the intrinsic mass of the object, and $1/D$ is a proportionality constant. The value of D is equal, by definition, to the average mass density in gravitational field.

Postulate (1) is meant to be applicable to any g-object, for instance, to the g-objects illustrated in figures 6 and 8. It states that the range of gravity of a g-object, which is defined by the volume of the g-object's gravitational field, is proportional to the total mass (m) of the g-object. The key notion incorporated in postulate (1) is that it is not the range of gravity in a single direction (a distance) but the entire "sum" of those ranges—that is, the entire volume of the gravitational field—that is directly proportional to the mass. Thus, the range of gravity represents a volumetric property of a g-object. It is represented by the volume of the spherical gravitational field in the case of the single-mass object shown in figure 5 and by the volume of the gravitational field depicted by the solid line in figure 6 in the case of g-object m_{1234}. Since D is assumed to be a constant, it follows that equation (1) does not change in space or time. With D being a constant, postulate (1) presents a simple, directly proportional relation between intrinsic mass and the range of gravity. I will show later, in section 21, that this is the only physically satisfactory relation.

Specifying the range of gravity in a single direction for a g-object would have no physical meaning, while specifying the "sum" of all ranges as the range of gravity of a g-object is rational and allows for identifying a well-defined property of a g-object, as will be discussed in section 19. The range of gravity in a single direction would not be invariant with respect to a location in the gravitational field of a g-object, whence it could not be a physical property. In contrast, the volume of gravitational field is the same regardless of the location in gravitational field. It can only change with a change in the total mass of a g-object.

It is important to note that the shape of the gravitational field of a g-object is, in general, unknown. (Exception in this regard is the special case of a single-mass object, for which the shape of gravitational field is approximately spherical.) Postulate (1) says nothing about that shape. One would need to introduce an additional mathematical assumption in order to define the shape of the gravitational field of a g-object.

To further appreciate the meaning of postulate (1), suppose that the single-mass object m_2 illustrated in figure 6 is moving vertically up (northward). Object m_2 is a constituent of g-object m_{1234}, which has the boundary of gravitational field delineated by the solid line. The range of gravity of the g-object in the northern direction—that is, the northern boundary of the g-object's gravitational field—will be moving primarily north (upward) as a result of the motion of object m_2, with some upper part of the boundary expanding east and west. The south-central part of the boundary of the g-object's gravitational field will primarily be moving northward as well. What postulate (1) adds here is that the total volume of the combined gravitational field $v_{1234} = (v_1 + v_2 + v_3 + v_4)$, which is proportional to the total mass of g-object m_{1234}, remains unchanged with the motion of object m_2—that is, with the deformation of the boundary of the gravitational field—as long as object m_2 remains a constituent of that g-object. This means that the total potential for gravitational interactions in all directions, represented by the volume of the gravitational field v_{1234}, remains constant as long as the total mass of the g-object remains unchanged.

Since postulate (1) incorporates the intrinsic mass, the range of potential gravitational interactions of a g-object does not depend on the motion of an observer or on the location of the g-object with respect to other g-objects. Put otherwise, the range of gravitational interactions of a g-object (i.e., the volume of the g-object's gravitational field) depends on its intrinsic mass only and is independent of the g-object's translational kinetic energy and gravitational potential energy. Nor does that range depend on the translational kinetic energies or the gravitational potential energies of the single-mass constituents of a g-object. That is because the volumes of the gravitational fields of the single-mass objects that form the g-object, which are represented by the dashed circles in figure 6, do not change with the motions of those objects, consistent with postulate (1). Therefore, the total volume of the g-object's gravitational field does not change either.[s] It is worth remembering that the intrinsic mass of an

---

[s] The premise that the intrinsic mass of a g-object does not depend on the translational kinetic energies of its single-mass constituents may seem objectionable. Looking at g-object m_{1234} (fig. 6), one might suppose that the translational kinetic energies of the constituent single-mass objects should be accounted for in determining the intrinsic mass (energy) of the g-object, since they represent the internal energies of the g-object as a whole similar to, for instance, the kinetic energies of atomic electrons in any material object. While this supposition may appear to be correct, the sum of those energies as measured

object includes its internal kinetic and binding energies—which means that it also includes the object's heat-energy content—whence a heated object is expected to have a larger range of gravitational interaction than the same object after cooling. An increase in gravitational effect with increasing temperature of an object has been confirmed by experiments (Steven Carlip[128]).

Since much of the remainder of this book is about the consequences of postulate (1), it will be beneficial to examine it further for a few minutes to fully appreciate its meaning. It has been introduced based on two underlying assumptions: first, a finite range of gravity was assumed, which is an implication of the infinite-time hypothesis and the only possible solution of the gravity paradox. Second, it was assumed that the finite range of gravity of a g-object, as defined by the volume of its gravitational field, is proportional to the total mass of a g-object. It will be shown throughout this book that postulate (1), while very simple, is also very powerful in terms of its consequences.

I have recently come across a paper by Robert Hooke, and found it rather amazing that Hooke, in the 1670s—that is, prior to the publication of Newton's *Principia*—might have thought of a finite range of gravity in terms similar to those outlined in the preceding paragraphs.[129] He stated, "For supposing all the fixt Stars as so many Suns, and each of them to have a Sphere of activity or expansion proportionate to their solidity and activity, and a bigger and brighter bodied Star to have a proportionate bigger space or expansion belonging to it." The expression "Sphere of activity or expansion proportionate to their solidity" sounds to me much like postulate (1), particularly when I look at the illustration of the gravitational field of the single-mass object presented in figure 5.

18 Speed of Gravity

The issue of the speed of gravitational interaction, referred to hereafter as the "speed of gravity," is critical to any cosmological model. The assumption that the speed of gravity equals the speed of light was a primary stimulus for Einstein to develop his law of gravitation, which he

---

at the CENTER OF MASS of the g-object is always equal to zero. Therefore, the kinetic energies of single-mass objects do not add to the intrinsic mass (energy) of the g-object formed by those objects.

called "general relativity." Einstein developed general relativity with the purpose of replacing Newton's law of gravitation, which implied that the speed of gravity was infinite. The same idea is implied by the statement of law III presented in section 7. It follows, therefore, that it will be constructive to present arguments supporting the concept of an infinite speed of gravity.

Law III, which states that two mass objects attract each other, is fundamental in the sense of the FLH. Consequently, it cannot have underlying mechanisms or principles, which implies that the speed of gravity is infinite. That is because there's nothing in the statement of law III that would imply the existence of a limit on the speed of gravity, just as there is nothing in Newton's law of gravitation that would imply the same. (Newton's law is summarized by equation (9) in section 22.) Furthermore, gravitational interaction *in itself*—which, as far as we know, comprises neither matter nor energy—cannot have a speed-control mechanism built into it. Since gravitational interaction in itself cannot interact with any matter or energy, there cannot be any conflicts that could limit the speed of gravity. Of course, conflicts may affect the speed of the motion of matter/energy that results from gravitational interactions but not the speed of interaction itself. (Conflicts affecting the executions of the fundamental laws of physics were discussed in section 10.)

According to special relativity, matter and energy cannot move faster than the speed of light, c. However, there is nothing in special relativity that would limit the speed of an interaction that is immaterial, that is, that comprises neither matter nor energy. To my knowledge, the only available "proof" showing that the speed of physical interaction, which can be thought of as the speed of information transmittal, has to be equal to or less than the speed of light, is based on the principle of causality. The conclusion that emerges from that proof is: if the speed of information transmittal were greater than the speed of light, there would exist inertial frames of reference in which an observed effect would precede its observed cause—that is, the principle of causality would be violated. The same idea is often stated in another way: if one could make the speed of information transmittal faster than c, then either the principle of causality or special relativity, or both would be proven wrong. However, there is something that has not been disclosed in those conclusions: the causality proof requires space-time to be a physical reality since a geometric space-time diagram has to be used in the proof, while that reality has never been established (confirmed). The existence of space-time as physical reality can only be *conjectured* based on the mathematical framework of Minkowski's space-time used to describe the behavior of matter and energy. However, such descriptions, no matter how successful, cannot prove space-time to have a physical meaning. Confirming that space-time is a physical reality would require a direct detection of it carried out independently of any physical law or theory. That has never been done. Thus, the proof of the limit on the speed of information transmittal can also be construed as an implication that the concept of space-time merely represents a mathematical framework. The claim that immaterial information (e.g., gravitational interaction) cannot travel faster than the speed of light *cannot* be concluded from special relativity since special relativity is based on the speed of light, which *is* material (light comprises energy).

To appreciate this discussion better, consider the following scenario: I am a man who threw a stone, which was the cause that broke a window, which was the effect. The cause preceded the effect. If the result of any thought experiment, carried out based on the principle of causality or otherwise, implies that the window was broken before the stone was thrown in an observer's frame of reference, that experiment has either been wrongly designed or carried out under wrong assumptions. Such a thought experiment would not prove that the speed of information transmittal cannot exceed the speed of light. It would merely prove that it was carried out or interpreted based on wrong assumptions.

One of the first physicists who considered Minkowski's geometric concept of space-time to be a mathematical instrument rather than a physical reality was Henri Poincaré[130], a coauthor of special relativity, from which the space-time concept was conceived by Minkowski. The principal author of special relativity, Einstein, was seemingly of a similar opinion. He is widely quoted as saying that "time and space are modes by which we think and not a condition in which we live." Today, more than one hundred years later, many accomplished physicists think the same thing. For instance, N. David Mermin plainly states: "...spacetime is an abstract four-dimensional mathematical continuum of points that approximately represent phenomena whose spatial and temporal extension we find it useful or necessary to ignore."[131] It appears that

Steven Weinberg had similar thoughts when he wrote with reference to a future theory of everything and to Minkowski's four-dimensional space: "I suspect...that what we are going to have is not so much a new view of space and time, but the de-emphasis of space and time. The spacetime coordinates may turn out to be just four out of the ten—or fifteen or twenty six or whatever it is—degrees of freedom that are needed to describe the theory."[132] And, if space-time represents a mathematical framework rather than physical reality, the causality proof becomes void.

Because the concept of space-time is critically important to the speed of gravity, I will now discuss the key aspect of the causality proof of the limit on the speed of information transfer, including the speed of gravity. As already stated, that key aspect is the physical reality of space-time. (If space-time is a physical reality indeed, then there is no question that the speed of physical interaction has to be equal to or less than the speed of light.) That reality could be confirmed if space-time is detected and measured independent of any physical law or theory. The following example illustrates such an independent act of measurement. In the ordinary three-dimensional space, an object travels along a trajectory—in general, along a curved line. The distance (Δd) between two close-by points along a trajectory is defined by the Pythagorean theorem, $(\Delta d)^2 = (\Delta x)^2 + (\Delta y)^2 + (\Delta z)^2$. In four-dimensional space-time (which has three spatial dimensions and one time dimension), an object travels along a world line. The "distance" between two close-by points (referred to as "events") along a world line, which is called a space-time interval (Δs), is defined by applying the Pythagorean theorem to the four dimensions of space-time as follows: $(\Delta s)^2 = (\Delta x)^2 + (\Delta y)^2 + (\Delta z)^2 - (c\Delta t)^2$. The three-dimensional space represents a physical entity for the purpose of formulating a physical law since distance (Δd) can be measured, and something that can be measured represents a physical entity. (Human space as a physical entity was discussed previously in section 13.) One can make such a measurement by applying a measuring device (a measuring stick) to an observed distance between two points. I know of no device that could be used to measure space-time interval Δs. The space-time interval can be neither observed nor measured. It can only be *calculated* from two independent measurements: the measurement of space and the measurement of time. Since a device that could be used to measure a space-time interval is not known, space-time cannot be confirmed to be a physical entity. As a result, the validity of the causality proof of a limit on the speed of gravity cannot be confirmed either.

In terms of experimental evidence, a debate on the speed of gravity recently took place following Edward Fomalont and Sergei Kopeikin's claim that they had measured the finite speed of gravity to be equal to the speed of light within a 20 percent margin of error (debate by Fomalont and Kopeikin[133], Clifford Will[134], Steven Carlip[135], Stuart Samuel,[136] and others). The debate led to the conclusion that the interpretation of Fomalont-Kopeikin experiment was incorrect. That means that a finite speed of gravity remained experimentally unconfirmed.

Another experimental approach that has been assumed to be capable of confirming that the speed of gravity is finite is the detection of GRAVITATIONAL WAVES. This is because the speed of gravity cannot be infinite if gravitational waves exist. It is important to realize, however, that the speed of gravity cannot be infinite only if space-time exists in physical sense. Then, a gravitational wave is supposed to be a "ripple" propagating through the fabric of space-time. Let me now comment on possible conclusions from the detection of gravitational waves without assuming, however, that space-time exists as a physical entity.

Numerous experiments designed to directly detect gravitational waves have been carried out over the last twenty-five years, with no success. Joel Weisberg and Joseph Taylor announced in 2004 a discovery of gravitational waves generated by the Hulse Taylor binary pulsar.[137] However, no gravitational waves were actually detected as a part of that discovery. Their existence was merely inferred from the rate of energy emission calculated based on the observations of decreasing PERIASTRON time over a thirty-year period.

Nonetheless, on February 11, 2016, B. P. Abbott et al. announced that gravitational waves were finally detected in the LIGO experiment.[138] According to the interpretation of the experiment results, a merger of two black holes resulted in the release of an enormous amount of gravitational radiation (energy), equivalent to about three sun masses, in a very short period of time. The speed of the motion of the released energy is, of course, expected to be finite, and equal to or less than the speed of light. That is required by special relativity: the speed of the motion of matter and energy has to be equal to or less than the speed of light. Therefore, the LIGO experiment *did not* prove that the speed of gravity—that is, the speed of a physical interaction comprising neither matter nor energy—is finite. The gravitational energy released as a result of the merger of the two black holes was converted from the gravitational potential and translational kinetic energies of the merging black holes. But, consistent with postulate (1), neither gravitational potential energy nor translational kinetic energy affect gravitational interactions as neither represents an intrinsic energy.[†] In other words, gravitational interaction, which represents a transfer of information that comprises neither matter nor energy, could not affect the results of the LIGO experiment, whence the speed of information transfer was not proven to be finite.

The most common argument used against Newton's theory of gravitation, which implies an infinite speed of gravity, is simply stated as: gravitational waves cannot exist in Newton's theory. Yet, even this argument appears to be incorrect. Bernard Schutz, for instance, demonstrates that gravitational radiation (waves) can be predicted from "Newtonian gravity and a little special relativity."[139]

19 Gravity (Gravitational Field)

According to postulate (1), the volume of a gravitational field (v) represents a uniquely defined property of mass, whence a new physical entity emerges. I will call it *Gravity* (spelled with a capital g) and denote it with symbol V. Its value equals the volume of the gravitational field (v). Postulate (1) can then be rewritten as

$$V = \frac{1}{D}m. \tag{2}$$

The symbol that identified the volume of a gravitational field (v) has thus been replaced with the symbol V to emphasize the physical meaning of Gravity. Equation (2) presents a definition of Gravity. In the remainder of this book, the word Gravity and the phrase "gravitational field" will be used interchangeably.

Gravity is *immaterial* as it comprises neither matter nor energy, just as in the case of Newton's "immaterial agent" that would mediate the

---

[†] Kinetic energy of an object depends on the speed of an observer's frame of reference. I find it unbelievable that gravitational interaction between the earth and a falling apple could depend on the speed of an observer passing the earth.

Gravitation 69

force of gravity. (See the quote of Newton in section 9.) However, this does not mean that Gravity does not represent a physical entity. To appreciate Gravity as a physical entity, consider its fundamental physical property, which was identified previously in section 9: if a grain of sand or any other mass object is inserted into a gravitational field, it will experience a gravitational interaction (a force of gravity). That effect can be observed and measured. It follows that Gravity can be detected in an experiment, which implies that it does exist as a physical entity. The other part of the proof that Gravity represents a physical entity rather than a purely geometric gadget can be derived from a closely related thought experiment. Let an electrostatically neutral grain of sand be inserted into a Gravity-free space. It will experience no gravitational force. I will conclude later, in section 37, that Gravity-free regions are, in fact, expected to exist in the universe. Gravity-free regions can exist because the range of gravity is finite, as implied by the GPU theory.

The result of the sand-grain test needs to be well understood. Let a grain of sand be inserted into the gravitational field of the earth at a location one hundred meters from the earth's surface. Let identical grain of sand be inserted into the gravitational field of the earth one hundred kilometers from the earth's surface. Both sand grains will be subject to gravitational interactions, however, those will be quantitatively different. That does not mean that the earth's gravitational field has a property with a magnitude that depends on location. The observed quantitative effects of gravitational interactions depend on the amount of conflicts with the executions of conflicting FLPs and/or conflicting matter/energy, and not on a location in gravitational field. All properties of gravitational field are exactly the same at each location in the field. In this regard, note that there seems to be only one property of gravitational field that can be discerned. It is the average mass density in gravitational field (D). This is contrary to the concept of a field normally used in physical laws. In the normal concept of a field, conflicts between the execution of an FLP and the interfering FLPs and/or interfering matter/energy are not recognized, and all contributions to the observed quantitative effects are lumped together according to a physical law. That leads to mathematical fields that have different field values, dependent on location. A gravitational field that has exactly the same property at each location is consistent with the FLH—that is, with the laws of nature that has been dehumanized.

As it is commonly used and understood, the word "gravity" has no quantitative meaning. Herein, Gravity is an exactly defined physical quantity. If the range of gravity were assumed to be infinite, which is the original Newton's proposal, the physically meaningful concept of gravity similar to that introduced herein could not exist. That is for any mass object would have an infinite range of gravity, whence the volume of its gravitational field would always be infinite regardless of the mass of the object (which would lead to the gravity paradox discussed in section 2).

Equation (2) and the mass-energy equivalence equation $E = mc^2$, where E and m denote intrinsic energy and intrinsic mass, respectively, yield

$$E = Dc^2V, \qquad (3)$$

which shows a relation between intrinsic energy and Gravity. It follows from equation (3) that energy and Gravity are inherently related in the sense that energy cannot exist without Gravity, and vice versa. This implies that energy causes a gravitational effect just as mass does, which also is a consequence of special relativity.[140] Therefore, massless objects, such as photons, are expected to be subject to gravitational interactions as long as such objects comprise intrinsic energy.

In developing special relativity, Einstein concluded that the emission of electromagnetic radiation causes the loss of mass in a mass object.[141] The amount of mass lost is quantized by the energies of emitted photons, which are finite. A finite loss of mass should result in a finite loss of a gravitational property of that object. According to equation (3), the loss of Gravity, which represents the loss of a gravitational property of a mass object, can now be quantized: an emitted photon carries a finite amount of Gravity (V_{ph}) directly proportional to its intrinsic energy (E_{ph}).

In section 29, I will submit that the kinetic and intrinsic energies of a photon are one and the same. That submission will lead to the premise of the gravitational interaction of a photon, which is a QUANTUM of pure energy. Consistent with the implication of equation (3), it is the intrinsic energy of any mass-energy object, including the photon, that causes a gravitational effect. Nonetheless, I will also show in section 29 that a photon would not be expected to behave in all respects in the same way that a massive particle with the same amount of Gravity would behave. The exception here is the ability of the (massless) photon to compensate for a change in its gravitational potential energy without changing its speed, which is an ability that the massive particle does not have. That ability of the photon leads to the conclusion that its speed is constant when subject to gravitation. Therefore, when subject to gravitation, the photon cannot be accelerated, as opposite to a massive object. This aspect of the photon's behavior will be discussed in section 32.

The notion of gravitational interactions of light with mass objects is old. It was first put forward by Newton's in *Optics*,[142] and has generally been accepted since the beginning of nineteenth century. Gravitational deflection of a light beam grazing the sun later became a leading test of general relativity. According to both general relativity and Newton's law of gravitation, the deflection of a light (radiation) beam passing next to a massive object does not depend on the frequency of the photons. The preceding discussion, on the other hand, suggests that the deflection of the passing beam (a gravitational effect) should depend on the photon frequencies (i.e., energies), as implied by equation (3). That dependence, however, is expected to be negligible and virtually unmeasurable as the photon has an *extremely small intrinsic energy in comparison with the mass of the interacting massive object* (e.g., the mass of the sun). This aspect of gravitational interactions will be discussed in section 22 in conjunction with the discussion of the finite range of gravity as well as in section 26 where the equivalence principle will be examined.

Gravity may also relate to the size of a photon. While we do not really know what the size of a photon is or even what the size of a photon means, we know from the definition of gravitational field that the volume of a photon cannot be larger than the volume of its gravitational field— that is, the maximum possible volume of a photon is equal to the volume of its gravitational field. If a photon is a fundamental particle in the sense of the FLH, then the assumption that the size of a photon is equal to the volume of its gravitational field would present a rational model of a photon, since it would eliminate the need for some internal forces or mechanisms that would be required to keep the size of a photon smaller than the volume of the field. To appreciate that model, examine figure 5. In the case of an ordinary mass object such as a stone, or a star, the internal mechanisms that may comprise gravitational, electromagnetic and strong nuclear forces, keep the size of the object smaller than the size of its gravitational field. Such objects can contain internal mechanisms as they are not fundamental particles. If one considers a photon to be a fundamental particle then, in accord with the FLH, it cannot contain any mechanisms. This suggests that the volume of the photon could be equal to the volume of its gravitational field.

20 Gravitational Interactions

The single-mass objects m_1 and m_3 illustrated in figure 6 are constituents of g-object m_{1234}. Those objects *directly* interact with each other since their individual gravitational fields are connected. Objects m_1 and m_4, on the other hand, do not directly interact with each other since their individual gravitational fields are separated, which means that the gravity ranges of m_1 and m_4 are too small for a direct gravitational interaction to occur between those two objects. Nonetheless, since their gravitational fields are within (that is, connected to) the gravitational field of the entire g-object m_{1234}, each of the objects m_1 and m_4 *indirectly* interact with all other single-mass objects that form the g-object. To appreciate this, consider a motion of object m_4. Because of its direct gravitational interaction with object m_3, which, in turn, directly interacts with objects m_1 and m_2, object m_4 indirectly interacts with those two objects. Any of the single-mass objects that form a g-object also interacts, indirectly, with all other single-mass objects via the interaction with the g-object m_{1234} as a whole.

Indirect gravitational interaction can be further explained by the example of two stars: star 1 and star 2. Let star 1 directly interact with star 2—that is, their gravitational fields are connected. Imagine now that star 1 is broken into a very large number of fragments (n) separated by some very small distances. According to the premise of the *infinite* range of gravity, each of those fragments directly interacts with star 2. In the physics of the infinite-in-time universe, which entails a *finite* range of gravity, that is not necessarily the case. Let each fragment of star 1 be so small that its gravitational field is not connected to that of star 2. One might conclude then that stars 1 and 2 no longer interact. That, however, would be an incorrect conclusion, as it would imply a complete loss of gravitational interaction between the two stars just because star 1 has been broken into a number of fragments separated by some miniscule distances. Moreover, that conclusion would also be wrong according to postulate (1), which requires the volume of a gravitational field to be proportional to the total mass of a g-object; in this case, to the total mass of the star 1 fragments. The gravitational field of the star 1 fragments would be practically the same as that of the original star 1, so each of the fragments would indirectly interact with star 2. Star 2 would interact with g-object $m_{1...n}$ much in the same way as it interacted with the original star 1. (Nature would not notice that star 1 was broken into fragments.)

Consider a lonely star situated far from the BULGE OF A GALAXY ("galaxy G") in which the great majority of the galaxy's total mass is concentrated. The lonely star is too far from any other star for a direct gravitational interaction to occur. Yet, actual observations of the motion of the star would show that the lonely star interacts with the entire mass of galaxy G. A description of that interaction can be approximated by the description of the lonely star interacting with a fictitious object that has the mass of the entire galaxy G and is located at the galaxy's center of gravity.

Consider the star 1 – star 2 scenario again. The gravity center of the fragments of star 1 represents the location of the fictitious object, which has a mass equal to the mass of all of star 1's fragments. From the lonely star – galaxy G scenario, we know that the indirect gravitational interaction between star 2 and the star 1 fragments can be approximated as a direct interaction of star 2 with the fictitious object representing the star 1's fragments. (Newton's gravity force equation could then be used to calculate the force of gravity between star 2 and the fictitious object.) But, that approximation would only be valid if the distances between the star 1's fragments are very small. This supposition is explained in the following paragraph.

Figure 9a shows a single-mass object m_0 that directly interacts with object m_1 before being broken into four single-mass objects (figs. 9b through 9e). Each of the four objects has a mass equal to $0.25 \times m_0$. The separation distances between the four objects in figure 9b are relatively small. Object m_1 would be expected to interact with the fictitious object located at the gravity center of the four single-mass objects in a way similar to the direct interaction with object m_0 shown in figure 9a (just like in the interaction in the 1 – star 2 scenario). In other words, the magnitude of the force of gravity acting on object m_1 would be expected

74 Infinite Universe

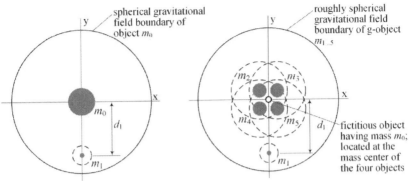

Fig. 9a Object m_1 is subject to direct gravitational interaction with object m_0.

Fig. 9b Object m_1 is subject to indirect interaction with the four single-mass objects via the fictitious object.

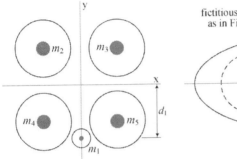

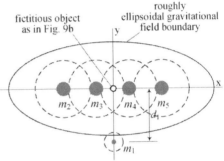

Fig. 9c Objects m_2 through m_5 are widely spread so that m_1 does not directly or indirectly interact with any of those objects.

Fig. 9d Object m_1 indirectly interacts with the four objects represented by the fictitious object; the strength of interaction expected to be near-zero.

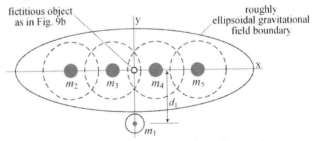

Fig. 9e Objects m_2 through m_5 are spread so widely that m_1 does not gravitationally interact, directly or indirectly, with any of those objects.

to be similar in figure 9a and figure 9b scenarios. Similar to figure 9b scenario, object m_1 shown in figure 9d interacts *indirectly* with the other four objects. As the four objects are spread farther and farther apart, the strength of the gravity force acting on object m_1 is expected to decrease.

In figure 9d, the combined gravitational field of objects m_2, m_3, m_4 and m_5 barely overlaps with the field of object m_1, whence the strength of the gravity force acting on object m_1 is expected to be very small. As the four objects are spread farther and farther apart, that strength will eventually become zero as the gravitational fields become separated (fig. 9e). This can be generalized by the following conclusion: as a number of objects, which indirectly interact with a single-mass object, are spread farther and farther apart, the strength of that interaction is expected to become weaker and weaker and to eventually become zero when the gravitational field of the single-mass object becomes disconnected. (The zero strength scenarios are illustrated in figs. 9c and 9e.)

Object m_1 is a constituent of g-object $m_{1...5}$, as illustrated in figures 9b and 9d. It indirectly interacts with the other single-mass objects (m_2, m_3, m_4 and m_5) as its gravitational field is connected to the gravitational field of g-object $m_{1...5}$. The force of gravity between those four objects and object m_1 can be calculated using an ordinary law of gravity (e.g., Newton's gravity force equation) if the four objects are replaced with a fictitious object. The fictitious object, with the mass equal to $m_2 + m_3 + m_4 + m_5$, would be located at distance d_1 from object m_1, as illustrated in figures 9b and 9d. As discussed in the preceding paragraph, however, the strength of the interaction with the fictitious object would be expected to decrease as the four objects are spread farther apart. This means that the mass of the fictitious object—which would have to be used in the calculation of the force of gravity between object m_1 and the fictitious object—would have to be taken to decrease with the spreading of the single-mass objects. The problem here is that the relation between the effective mass of a fictitious object and the spread of its single-mass constituents is not known. ("Effective mass" is that needed to properly calculate the approximate strength of the force of gravity). It is only where the mass objects that form a fictitious object are very close to each other, the mass of the fictitious object can be taken to be approximately equal to the sum of the masses of the four objects.

The conclusion on the decrease of the strength of gravitational interactions with the spreading of masses, which was drawn considering the single-mass object m_1, applies to any g-object. As outside masses—

that is, the masses that interact with a mass object or a g-object indirectly only—are spread farther and farther apart, the strength of the interaction becomes weaker and weaker and then negligible in comparison with the strength of direct interactions between the g-object's constituents. (This implication applies to typical conditions. A nontypical condition would exist where the gravitational fields of two constituents of a g-object are barely connected and the strength of the direct gravitational interaction between them is negligible as well.)

To further appreciate the foregoing conclusion on the decrease of the strength of gravitational interaction with the spreading of masses, consider the sun. The sun indirectly interacts with the entire Milky Way galaxy. If all of the Milky Way's stars were spread farther and farther apart, the strength of the indirect interaction between the sun and the other stars would become weaker and weaker. Eventually, that strength would become negligible or nil.

Another fundamental problem with a mathematical description (i.e., a physical law) of gravitational interactions can be concluded from an examination of figures 7 and 8. As figure 7 shows, single-mass object m_5 interacts with g-object m_{1234}. In reality, however, g-object m_{1234} does not exist anymore. It was transformed into g-object m_{12345} (see fig. 8) when the gravitational fields of m_5 and m_{1234} became connected. Thus, object m_5, by interacting with the fictitious object formed by all five single-mass objects, interacts indirectly with both the other four objects as well as with itself. (In the previous discussion of figure 9, no consideration was given to the indirect interaction of object m_1 with itself.) The indirect interaction of an object with itself leads to a circular reference problem, which has no mathematical solution. This is the same problem as the well-known n-body problem associated with Newton's law of gravitation. As the preceding discussion shows, however, this is not necessarily a mathematical problem only. It is a problem that can be identified, in physical terms, from the consideration of a finite range of gravity.

21 Gravity-Conservation Law

The gravity-conservation law, stated as law X in section 7, "Gravity cannot be created or destroyed", results from the supposition that Gravity is a physical quantity (section 19). This is for according to the general statement of the conservation laws suggested in section 8, "nature has no hidden sources of, or storage bins for, physical quantities", Gravity has to be conserved.

Another way to look at the gravity conservation law is to realize that the notion underlying postulate (1) contains the following symmetry: for a given mass of a g-object, irrespective of the point in time; irrespective of how many single-mass objects form the g-object; how the mass is distributed between those objects; where the objects are located; how the objects are moving; what are their mass densities, internal structures, or chemical compositions; the amount of Gravity, which equals the volume of gravitational field of the g-object, remains the same. That symmetry implies the gravity-conservation law.

Owing to equation (3), the gravity-conservation law must also hold in the presence of mass or energy creation. For instance, in the collision of two gamma-ray photons in which an electron-positron pair is created, the gravity-conservation law must be satisfied alongside the charge, mass-energy, and momentum conservation laws. That means that the two photons had, prior to the collision, exactly the same amount of Gravity as the electron-positron pair had following the collision

Consider two mass objects in motion toward each other. Their gravitational fields are just about to connect, as illustrated in figure 10. Figure 11 illustrates the volume of the combined gravitational field for the scenario after the gravitational fields of the two objects are connected and the objects gravitationally interact, which is a direct interaction. Due to the gravity conservation law, the total volume of the two initial gravitational fields $(V_1 + V_2)$ must be equal to the volume of the combined gravitational field V_{1+2}, as indicated in figure 11. It follows that the gravity conservation law also implies that the relation between Gravity (V) and mass (m) must be *directly* proportional. To verify that statement, assume that the relation between V (or v) and m is not directly proportional and is given by

$$V = f(m) \times m. \tag{4}$$

Then, owing to the gravity-conservation law, the sum

$$f(m_1) \times m_1 + f(m_2) \times m_2, \tag{5}$$

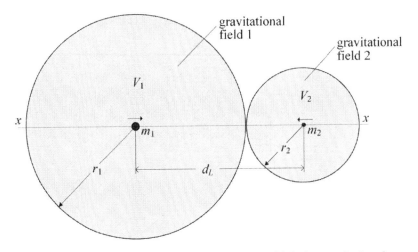

Fig. 10 Distance d_L between objects m_1 and m_2 at which the gravitational interaction will start or end, depending whether the two objects are moving towards or away from each other.

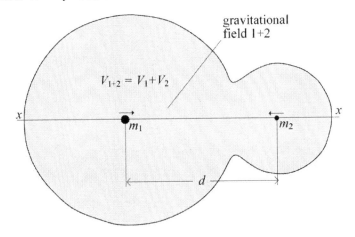

Fig. 11 At a distance $d < d_L$ between the two mass objects m_1 and m_2, the objects are subject to direct gravitational interaction.

which corresponds to the total amount of Gravity $(V_1 + V_2)$ before the gravitational fields connect (fig. 10), must be equal to

$$f(m_1 + m_2) \times (m_1 + m_2), \tag{6}$$

which is the total amount of Gravity (V_{1+2}) after the two gravitational fields connect (fig. 11). Therefore

$$f(m_1) \times m_1 + f(m_2) \times m_2 = f(m_1 + m_2) \times (m_1 + m_2). \qquad (7)$$

This is possible only if

$$f(m_1) = f(m_2) = f(m_1 + m_2) = \text{const.}, \qquad (8)$$

which means that the $f(m)$ included in equation (4) must be a constant. It follows that the relation between V and m, or between v and m, has to be directly proportional, which has been accounted for in the statement of postulate (1).

22 Gravity-Force Equation

In this section, I will derive a gravity-force equation from the consideration of the simplicity-of-execution principle and postulate (1). That equation is intended to describe a direct gravitational interaction between two mass objects. Its mathematical form will turn out to be the same as Newton's law of universal gravitation, except for a simple, first-approximation modification necessary to account for a finite range of gravity. I will not suggest that law to be a fundamental law of nature. It is meant to be a physical law (a law that exists in the human mind only).

Newton derived his law of gravitation from observations and experiments. Mathematically, Newton's law is expressed as

$$F_g^a = G \frac{m_1 m_2}{d^2}, \qquad (9)$$

where F_g^a is the force of gravity, m_1 and m_2 are the masses of two gravitationally interacting single-mass objects, d is the distance between the gravity centers of those objects, and G is a gravitational constant. The superscript a is included in equation (9) to indicate that the force of gravity calculated from it is thought to represent an *apparent* force of gravity (F_g^a) as opposite to the *true* force of gravity (F_g), the importance of which will be discussed in sections 23-25.

The gravitational interaction between two single-mass objects is expected to be affected by conflicts with the executions of other FLPs and/or conflicts with other matter/energy objects. As an example, in the earth-sun gravitational interaction, conflicts could result from indirect gravitational interactions of the sun and the earth with all stars in the Virgo supercluster—that is, from interactions with distant masses. The earth-sun gravitational interaction would also be affected by a conflict with the execution of law II (mass resists a change to its state of motion) and some other FLPs (e.g., the SLT, as I will discuss in section 25). Thus, unless such conflicts have a negligible effect on gravitational interaction between two objects, it has to be assumed that equation (9) implicitly accounts for those conflicts. That is because of the proven high degree of accuracy of Newton's gravity-force equation under typical experimental conditions. With regard to the conflicts that may affect gravitational interactions between two objects, recall from the discussion in section 20 that the influence of distant masses on local interactions is expected to be negligible.

It is impossible to conclude from equation (9) that one of the two gravitationally interacting objects contributes to the force of gravity more than the other.[u] Neither such a conclusion can be drawn from any experiments or observations. That is consistent with the simplicity-of-execution principle (section 10), which states that the execution of a fundamental law is nonquantitative. Specifically, that principle states that there cannot be any principles, weight scales or decision-making mechanisms built into an FLP (law III in this case) that would allow for enforcing quantitatively different actions dependent on the masses of the interacting objects. Put otherwise, the simplicity-of-execution principle suggests that the contributions of two mass objects to the apparent force of gravity between them should be taken equal for the purpose of deriving a physical law of gravitation. I will call that suggestion "implication (a)."

Consider the two gravitationally interacting objects m_1 and m_2 illustrated in figure 12, and assume that distance d_0 between the objects is the same in all of the considered scenarios. Let the masses of the objects be the same in the first three scenarios (i.e., $m_1 = m_3 = m_5$ and

---

[u] It follows from equation (9) that if the magnitude of mass m_1 increases twice, the force of gravity increases twice. Also, if the magnitude of mass m_2 increases twice, the force of gravity increases twice. Therefore, it is impossible to say that one mass contributes to the force of gravity more than the other one.

Gravitation

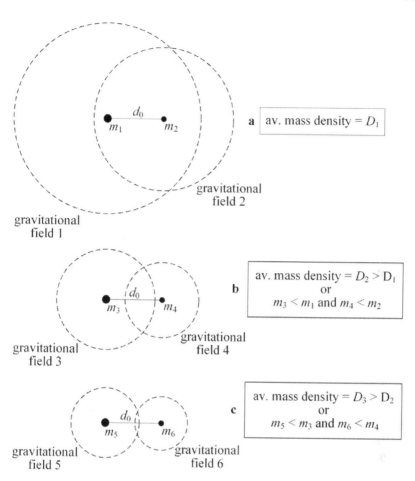

Fig. 12 Expected decrease in the strength of gravitational interactions between two objects with either an increase in the average mass density in gravitational field or a decrease in the masses of the objects (combined gravitational fields, such as that shown in figure 11, are not shown).

and $m_2 = m_4 = m_6$), while the average mass density in the gravitational field increases from figure 12a to figure 12c ($D_1 < D_2 < D_3$). Consistent with postulate (1), the radii of the gravitational fields of the two objects decrease with an increase in the average mass density D. (For a single-mass object, the range of gravity in any direction equals the radius of its gravitational field, which is calculated as $[(3m)/(4\pi D)]^{1/3}$.) In figure 12c, the gravitational fields of the two objects have decreased to the point that they are barely connected. That suggests a very low strength of gravitational interaction between objects m_5 and m_6. It follows that the strength of the gravitational interaction (the force of gravity) between the two objects is expected to decrease as density D increases. With a slightly higher density D_3, that force would become zero as the two gravitational fields are separated. Therefore, an inverse proportionality of the force of gravity to the average mass density in the gravitational field is expected. I will call this expectation "implication (b)."

It needs to be noted that implication (b) presents just an assumption made based on the supposition that the magnitude of the force of gravity becomes near zero when two gravitational fields are barely connected. Note also that the individual gravitational fields shown in figure 12 cannot actually be assumed to "overlap". That is because postulate (1) requires the total volume of a g-object's gravitational field to be directly proportional to the total mass contained in the system, as shown in figure 11. Figure 12 does not show the gravitational fields of the two object systems. It shows individual gravitational fields of the two mass objects.

Assume now that the average mass density in gravitational field D remains the same, while the masses of the two objects decrease from figure 12a to figure 12c as follows: $m_1 > m_3 > m_5$ and $m_2 > m_4 > m_6$. Consequently, the gravitational fields of those objects decrease as well. Qualitatively, the effect of the increases in the average mass density is the same as the effect of the decreases in the masses of the interacting objects. Again, the gravitational fields in figure 12c are barely connected, which suggests a very low force of gravity compared with those in the figure 12a and figure 12b scenarios. This suggests that the strength of the gravitational interaction decreases with a decrease in the mass of an interacting object—which, of course, is a well-known experimental fact. I will refer to this suggestion as "implication (c)." In the remainder of this section, my goal is to construct a physical law of gravity that is consistent with implications (a), (b), and (c).

There are four variables in postulate (1) that are intrinsic to the gravitational interaction between two single-mass objects, as illustrated in figures 10 and 11. Those include masses m_1 and m_2 and Gravities V_1 ($= v_1$) and V_2 ($= v_2$). To satisfy implications (a), (b), and (c), those four variables can be arranged into two terms—namely, $1/2 \times m_1 V_2$ and $1/2 \times m_2 V_1$, with the apparent force of gravity proportional to their sum as follows:

$$F_g^a \propto \frac{1}{2}m_1 V_2 + \frac{1}{2}m_2 V_1 \left(= \frac{1}{D} m_1 m_2\right). \tag{10}$$

Those two terms represent the contributions of single-mass objects m_1 and m_2 to the apparent force of gravity. In accord with postulate (1), each of those terms always has the same value as the other (they are perfectly symmetrical), whence implication (a) is satisfied. It follows from postulate (1) that $V_1 = m_1/D$ and $V_2 = m_2/D$, which means that the apparent force of gravity is in a reverse proportion to the average mass density in gravitational field, as assumed in proportionality (10). Therefore, implication (b) is satisfied. It further follows that the apparent force of gravity is proportional to either of the two masses (m_1 or m_2), which means that implication (c) is satisfied as well.

The terms $m_1 V_2$ and $m_2 V_1$ do not seem to have any physical meaning, which is analogous to the lack of physical meaning of the product $m_1 \times m_2$ in Newton's law of gravitation (9). Introducing those terms represents a mathematical assumption only, which has been made to make proportionality (10) consistent with implications (a), (b), and (c). Implication (a) results from the simplicity-of-execution principle, which, in turn, is a consequence of the FLH. Implications (b) and (c) result from the consideration of postulate (1), which is a consequence of a finite range of gravity implied by the infinite-time hypothesis. It follows that the FLH and the infinite-time hypothesis represent the foundations of proportionality (10).

To derive a physical law of gravitation, space and time have to be incorporated into proportionality (10). Space is taken to be the distance (d) between two interacting objects. First, let me average each of the four variables included in proportionality (10) over space

$$F_g^a \propto \frac{1}{2}\frac{m_1}{d}\frac{V_2}{d} + \frac{1}{2}\frac{m_2}{d}\frac{V_1}{d}. \tag{11}$$

Second, let me average each of those four variables over a time period t_p

$$F_g^a = \frac{1}{2}\frac{m_1}{t_p d}\frac{V_2}{t_p d} + \frac{1}{2}\frac{m_2}{t_p d}\frac{V_1}{t_p d}. \tag{12}$$

Using equations (2) and (12) one arrives at

$$F_g^a = \frac{1}{KD}\frac{m_1 m_2}{d^2}, \tag{13}$$

where $K \equiv t_p^2$. Equation (13) becomes identical to Newton's law of gravitation (9) if constant $1/KD$ is taken to be the gravitational constant G. Thus, by simple averaging of the four basic variables introduced in postulate (1), one arrives at Newton's law of gravitation, starting with the basic premise of equal contributions to the apparent force of gravity by two interacting mass objects and the two implications of postulate (1). Note that the averaging over passing time, as opposed to averaging over a period of time, would not be rational, which can be appreciated by considering the force of gravity between an apple hanging from a tree and the earth. As time passes, that force does not change (until the apple starts falling). So averaging over passing time would make no sense. While the physical meaning of space, which is represented by distance d, is obvious, the physical meaning of time period t_p is a priori unknown. It would be expected to represent a time period characteristic of the gravitational interactions between two objects. A possible physical meaning of t_p will be discussed at the end of this section.

The foregoing derivation of equation (13) is not based on some well-established approach to formulating physical laws from theoretical considerations. In particular, it is not clear why averaging of the four variables over distance and over a time period is required to derive a physical law. (Another average incorporated in (13) is the average mass density in gravitational field D.) It could be that the descriptions of some physical phenomena can be well approximated by using averages of basic variables. Regardless, my primary goal in presenting a theoretical derivation of equation (13) was to show that the implication of the equal contributions of two interacting mass objects to the force of gravity is in very good, if approximate, agreement with the results of all experiments and observations against which Newton's equation (9) has been tested. This suggests that the executions of FLPs are indeed nonquantitative, which was a conclusion drawn from the simplicity-of-execution principle in section 10. The other goal was to conclude that the magnitude of the apparent force of gravity is in reverse proportion to the average mass density D.

In respect to the averaging procedure used to derive equation (13), I note that the process of averaging quantities over a distance and a period of time is rather common in formulating physical laws. Typically, where a flux is proportional to the gradient between two quantities, averaging those quantities over a distance and a time period is used to formulate a physical law. A gradient represents the difference between the averages of two quantities (Q_1 and Q_2) over a distance d: $\nabla Q = Q_2/d - Q_1/d$. A constant period of time (Δt) is incorporated into the equation's material constant (C), similar to incorporating time period t_p in equation (12). The typical form of such a physical law is FLUX = $C \nabla Q$. Darcy's law of water flow through porous media, Fick's law of diffusion, and Fourier's law of heat conductance are examples of such physical laws.

According to the premise of a finite range of gravity, equation (13) is incorrect, since it implies that gravitational interaction occurs at any distance (d) between two mass objects. According to that premise, however, the force of gravity should become zero at distances $d \geq d_L$. (Distance d_L is defined in fig. 10.) To account for that condition, equation (13) can be modified by introducing, as a first approximation, the correction factor equal to $1 - d/d_L$

$$F_g^a = \frac{1}{KD} \frac{m_1 m_2}{d^2} \left(1 - \frac{d}{d_L}\right). \tag{14}$$

At the distance $d = d_L$, the force of gravity becomes zero. At a distance $d > d_L$, there is no gravitational interaction between two mass objects, and equation (14) does not apply (the force of gravity cannot be negative according to the fundamental law of gravitation III). There are no available data that would allow one to verify the suitability of the mathematical form of the correction factor included in equation (14). That is because there are no sufficiently accurate experimental data on the force of gravity between two mass objects located at relatively large distances from each other, for instance, between an Oort cloud comet and the sun. At typical experimental/observational distances (e.g., the sun-planet distances) such data are available, but the correction factor would be too close to unity to allow for verifying its suitability. One will be able to better appreciate the "too close to unity" stipulation after the ranges of gravitational interactions are discussed in section 27.

Equation (14) could still be seen as incorrect as it implies that the force of gravity becomes infinite at $d = 0$, which would lead to another gravity paradox: gravitational interactions between finite masses results in the generation of an infinite force. In this regard, one needs to realize that equation (14) is applicable only to situations in which two interacting single-mass objects can be represented by point masses. This results from proportionality (10), which does not depend on the sizes of interacting objects. To appreciate this point, consider the gravitational interaction between two objects comprising fluids. As long as the two objects are separated, they can be represented by point masses, and proportionality (10) and equation (14) are applicable. If the two objects connect, they can no longer be represented by two point masses since only one object and one point mass exist. Consequently, proportionality (10) and the resulting equation (14) are no longer applicable. The gravitational interactions inside a single object formed by two originally separated objects, would have to be described by a different physical law of gravitation. This notion is analogous to the case of electrostatic interactions, as explained by John A. Wheeler:[143]

> Near a proton, the electric field becomes stronger and stronger the closer you get to the particle. If the proton were a true point, the field would become infinitely strong when you reach the proton. But in fact that doesn't happen. The proton has a finite size. It has structure. To be sure, everywhere outside the proton, the electric field has the same strength as if the proton were a point. But when you penetrate inside the proton, you find that the electric field is merely very large, not infinite. The proton's size and structure put a lid on the strength of the field.

Wheeler refers to a proton that "has a finite size and a structure." From the perspective of gravitational interactions, mass objects have structures as well. Mass objects comprise atoms and molecules. Interactions between atoms and molecules involve electromagnetic and gravitational energies. The internal motions (vibrations) of atoms and molecules involve kinetic energies. The forces associated with those energies lead to conflicts with the execution of the fundamental law of gravitation. What needs to be appreciated here is that those conflicts would be quantitatively different before and after two gravitationally interacting mass objects connect. In Wheeler's words, "The proton's size and structure put a lid on the strength of the field." The same applies to gravitational interactions: the structures of the interacting mass objects—that is, atoms and molecules, and their internal interactions—result in conflicts, which "put a lid" on the strength of the force of gravity.[v]

If one applies Newton's second law of motion ($F = ma$) to equation (14) in the usual way (i.e., $F_g^a = F$), one finds the equation for the acceleration (a_2) of object m_2 toward object m_1, as measured in the m_1 reference system, to be

$$a_2 = \frac{1}{KD} \frac{m_1}{d^2} \left(\frac{d_L - d}{d_L} \right) \qquad (15)$$

or, using the definition of d_L illustrated in figure 10 and the sphere volume formula

$$a_2 = \frac{1}{KD} \frac{m_1}{d^2} \left[1 - \frac{d}{\left(\frac{3}{4D\pi}\right)^{\frac{1}{3}} \left(m_1^{\frac{1}{3}} + m_2^{\frac{1}{3}}\right)} \right]. \qquad (16)$$

Equation (16) illustrates an important implication of a finite range of gravity. It shows that for any object m_2 that has a *very small mass* relative to mass m_1, its acceleration (a_2) would be *approximately* the same. (According to Newton's law of gravitation, for a given mass m_1, the acceleration of object m_2 is *exactly* the same regardless of its mass.) To appreciate this inference, let object m_2 be in a free fall toward object m_1 along distance d_L, and assume that mass m_2 is very small in comparison to mass m_1. d_L is the maximum possible free-fall distance, which is illustrated in figure 13. For objects m_2 that have very small

---

[v] This also is similar to Newton's shell theorem, which applies to objects inside and outside of a hollow spherical mass object (a hollow ball). Newton demonstrated that a hollow ball exerts a force of gravity on an outside object as if it were represented by its point mass (i.e., as if the entire mass of the hollow ball were concentrated at the center of the ball). However, no net gravitational force would be exerted by a hollow ball on any object inside the ball.

masses relative to mass m_1, distances d_L are very similar, whence their accelerations and the maximum free-fall times (t_L) are very similar as well. This can be demonstrated using the kinematic relation between distance (d), acceleration (a), initial velocity (v_i), and time (t): $d = v_i t + 0.5 \times at^2$. That relation and equation (16) were used to calculate the following free-fall times t_L.

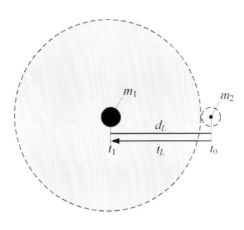

Fig. 13 Maximum free-fall distance and time for a (very small) mass object m_2.

Assuming $m_2 = 1$ kg, the values of t_L for the masses of the moon (3.45×10^{22} kg), the earth (5.97×10^{24} kg) and the sun (1.99×10^{30} kg) were calculated to be approximately 4.05×10^7 years, with a maximum difference (between the sun scenario and the moon scenario) of 1.5 years. To better appreciate the meaning of a "very small mass," assume that $m_2 = 1{,}000$ kg and m_1 to be the mass of the earth. For this scenario, calculations show that the free-fall time would be practically the same as in the $m_2 = 1$ kg case: it would be longer by about 3 years only, relative to the total free-fall time of about 4.05×10^7 years.

It can be shown that as mass m_2 increases while mass m_1 is kept constant, the time of the free fall over distance d_L increases. It follows that the time period $t_L \approx 4.05 \times 10^7$ years is the shortest possible free-fall time for any two objects that were initially separated by distance d_L. Therefore, the time period $t_L \approx 4.05 \times 10^7$ years can be considered to be a characteristic (a property) of the gravitational interactions between two mass objects. It needs to be emphasized that the value of period t_L was calculated using the specific mathematical form of the correction factor that was introduced, as a first approximation, in equation (14).

For relatively short distances between gravitationally interacting mass objects, the value of the correction factor equals approximately 1.0. Assuming this value, equations (9) and (14) can then be compared and the relation between the constants included in those equations can be derived, which yields $t_p^2 = 1/(GD)$. From this relation and the results of free-fall computations for objects with various masses, it follows that

$$t_p = 0.175 \times t_L. \tag{17}$$

Relation (17) does not depend on mass density D and is valid, to a very high degree of accuracy, as long as the mass of one of the two interacting objects is very small relative to the mass of the other object. The meaning of the 0.175 multiplier is not unclear. It could be the result of an approximation incorporated in equation (14), a calculation error, or another unaccounted for factor. Regardless, relation (17) suggests that constant t_p, which was introduced in equation (12) in conjunction with the averaging of the variables over a time period, might have a physical meaning (which I cannot discern). That suggestion arises from the fact that the time period t_L has an obvious physical meaning. If constant t_p does have a physical meaning indeed, then Newton's gravitational constant $G = 1/(t_p^2 \times D)$ would acquire a physical meaning as well.

23 Gravity versus Electrostatic Force

In this section, I will discuss the huge ratio of the electrostatic force to the gravity force generated between two elementary particles. As any fundamental law of physics has to be nonquantitative (see section 10), one can expect that the ratio of the two forces should be explainable in terms of the conflicts that affect, in different ways, the execution of the laws of electrostatic interaction (laws IV and V) and the execution of the law of gravitational interaction (law III). I will clarify this expectation shortly. In sections 24 and 25, I will suggest possible explanation of the huge difference between the strengths of the electrostatic and gravity forces.

The approximate strength of the electrostatic force (F_e) between two charged particles q_1 and q_2 can be calculated from Coulomb's equation

$$F_e = k_e \frac{q_1 q_2}{d^2}, \tag{18}$$

where q_1 and q_2 are the magnitudes of charges, d is the distance between the charged particles, and k_e is Coulomb's constant. The electrostatic force acting between a proton (p^+) and an electron (e^-) is illustrated in figure 14a. Note that in accord with equation (18), the magnitude of the electrostatic force does not depend on the masses of the interacting particles—that is, it does not depend on the force of gravity between the two particles. It is important to keep in mind, however, that the particles do gravitationally interact in addition to their electrostatic interaction.

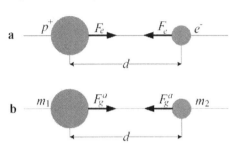

Fig. 14 Electrostatic vs. gravity force – interacting charges and masses are subject to the same (non-quantitative) actions.

The magnitude of the force of gravity between two single-mass objects can be calculated from equation (13). For simplicity, I will now neglect the correction factor introduced in equation (14). This will not affect any of the following conclusions. The apparent force of gravity force acting between two mass objects is illustrated in figure 14b. As pointed out in section 22, the superscript a included in the symbol of the force of gravity (F_g^a) indicates that that force is thought to be an apparent force of gravity as opposite to the true force of gravity (F_g).

Assume that objects m_1 and m_2 shown in figure 14b have masses equal to the masses of a proton and an electron, respectively. From equations (13) and (18), the ratio of the electrostatic force (F_e) to the apparent force of gravity (F_g^a) for the proton-electron and the m_1-m_2 systems can be calculated as

$$\frac{F_e}{F_g^a} = 2.25 \times 10^{39}. \tag{19}$$

This ratio applies specifically to the interaction of a proton and an electron. For different objects with the same charge values, the ratio F_e/F_g^a would be different. For instance, if the mass of the proton were replaced with the mass of a sodium ion, that ratio would decrease to 9.91×10^{37}. If the two particles had masses of 1.85×10^{-9} kg each and the charge magnitudes of a proton and an electron, the ratio F_e/F_g^a would be equal to 1.0. Feynman commented on the huge ratio given in equation (19) as, "It is hard to believe that they could both [i.e., the two forces] have the same origin."[144] Owing to the simplicity-of-execution principle, that comment can now be rephrased: It is hard to believe in such a huge ratio since the electrostatic and gravity forces do, in fact, have analogous origins. To appreciate that analogy, examine the statements of laws III and IV presented in section 7, while keeping in mind that the executions of both of those laws are nonquantitative, as discussed in section 10.

The gravitational and electrostatic actions enforced by nature when executing law III (gravitational interaction) and law IV (electrostatic interaction) are both nonquantitative and thus exactly the same: the proton and the electron are "pushed" toward each other in exactly the same way that the two electrically neutral mass objects m_1 and m_2 are pushed toward each other. Therefore, if one accounts for the executions of laws III and IV only, one can reasonably expect the ratio F_e/F_g^a to be

$$\frac{F_e}{F_g} = 1. \tag{20}$$

Ratio (19), which has been well established from observations and experiments, is in an enormous disagreement with ratio (20). Since the prediction expressed by ratio (20) is based on the simplicity-of-execution principle, which is a direct consequence of the FLH, one might judge the FLH to be inadequate because of the enormous discrepancy between ratio (19) and ratio (20). If, on the other hand, one believes that the FLH should hold, then there must be one or more conflicts with the execution of other FLPs and/or conflicts with other matter or energy that differently affect the proton-electron and the m_1-m_2 systems, which leads to such a discrepancy.

24 The Influence of Distant Masses

In this section, I will examine Mach's principle. This examination will guide the search for the explanation of the discrepancy between ratio (19) and ratio (20), which was identified in the preceding section.

The force of gravity acting between two mass objects results from the execution of the fundamental law of gravitation (law III in section 7) and from one or more conflicts with the execution of other FLPs and/or conflicts with other matter or energy. The effects of those conflicts have to be different from those of the conflicts that affect the execution of law IV illustrated in figure 14a, as discussed in section 23. Possible conflicts that could affect the execution of law III in a system such as the one illustrated in figure 14b include:

Possibility 1   The influence of distant masses, which is mediated by a gravitational field, results in a conflict with the execution of law III at a local level. A familiar analogy here is Mach's principle, which states that a law of gravitation is influenced by distant masses.[w]

Possibility 2   The influence of the local execution of one or more interfering FLPs. A familiar analogy here is general relativity, in which the effect of the force of inertia results from a local property of space-time.

As noted, the premise concerning the influence of distant masses on a local law of gravitation relates to the principle put forward by Ernst Mach in 1893.[145] In today's physics, the common statement of Mach's principle is: the relative motions of distant masses influence a local law of gravitation. An example of the applicability of Mach's principle is a spinning skater whose arms are pulled away as the result of gravitational interactions of the skater with distant masses. Mach's idea has been incorporated into some models of gravitation conceived, for instance, by Dennis Sciama[146] and by Carl Brans and Robert Dicke.[147] The concept of the force of inertia generated by gravitational interactions with distant masses is schematically illustrated in figure 15, which reveals the role of the acceleration of distant masses. In the following, I will replace the term "distant masses" with "outside masses," which better reflects the premise of a finite range of gravity. "Outside masses" are the masses outside of a g-object (i.e., outside of a local gravitational system) that do not directly interact with any constituents of the g-object.

---

[w] In section 7, the law of inertia was identified as an FLP. Consequently, it cannot have any underlying mechanisms. In this regard, Mach suggested that the force of inertia is caused by gravitational interactions with distant masses. If so, the law of inertia would incorporate a mechanism (interaction with distant masses) and could not hold the status of an FLP.

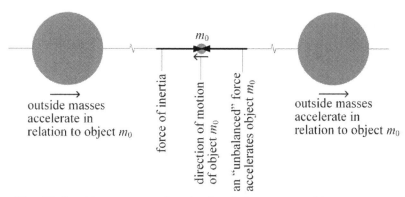

Fig. 15 Outside masses accelerating in relation to mass object m_0 generate the force of inertia acting on it (Mach's principle).

I will begin the explanation of the ratio (19) versus ratio (20) discrepancy by considering possibility 1 and examine the gravitational interaction between two mass objects illustrated in figure 16a. The masses of objects m_1 and m_2 are assumed to be equal to the masses of a proton and an electron, respectively. The potential influence of outside masses, if any, is first neglected. In accordance with ratio (20), the measured force of gravity should be equal to the electrostatic force if all conflicts with the execution of law III (the law of gravitational interactions) and with the execution of law IV (the law of electrostatic interactions) result in the same quantitative effects. Yet the actual force of gravity between objects m_1 and m_2 is well known to be orders of magnitude smaller than the corresponding electrostatic force, as indicated by ratio (19). It follows that to resolve the ratio (19) versus

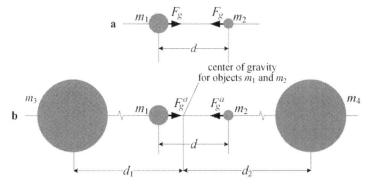

Fig. 16 Influence of distant masses on the apparent force of gravity.

ratio (20) discrepancy, either the influence of outside masses cannot be neglected, or there is another interaction that results in a conflict with the execution of law III that has not been accounted for.

Let g-objects m_3 and m_4, both of which have very large masses, be added to the m_1-m_2 system, as illustrated in figure 16b. Those two g-objects are located at the same distance ($d_1 = d_2$) from the center of gravity of objects m_1 and m_2. Distances d_1 and d_2 are such that each of the g-objects m_3 and m_4 interacts with both objects m_1 and m_2. The magnitude of the force of gravity between objects m_1 and m_2 would be affected by the gravitational interactions with outside masses, since g-objects m_3 and m_4, which are meant to represent outside masses, would tend to "pull apart" the local objects m_1 and m_2. Thus, the gravitational interactions with g-objects m_3 and m_4—that is, the influence of outside masses—would cause a reduction in the magnitude of the true force of gravity between objects m_1 and m_2. As a result, the discrepancy between ratio (19) and ratio (20) could be explained by the gravitational influence of outside masses on a local gravitational system.

While the above explanation may seem reasonable and attractive, there is a problem with it: the actual influence of outside masses on a local system is expected to be negligible. That was concluded previously in section 20. With respect to the scenario illustrated in figure 16b, an implicit while incorrect assumption was made: it was assumed that g-objects m_3 and m_4 interact directly with objects m_1 and m_2—that is, with the constituents of a local gravitational system. Direct gravitational interaction of a constituent of a local gravitational system (m_1 or m_2 in this case) with outside masses is not possible, in accord with the definition of a g-object (stated in section 17), which represents a local gravitational system. For instance, two stars that directly interact with each other in the Milky Way's gravitational field would not *directly* interact with any star in the Andromeda galaxy, or any other star in the Virgo supercluster,[x] all of which represent outside masses. The two stars would interact with outside masses indirectly only. And, as was

---

[x] It is of interest to note that Virgo supercluster, to which the Local Group and thus the Milky Way belong, was recently downgraded to the status of an appendage to the much larger astronomic structure identified as the Laniakea supercluster (R. B. Tully, H. Courtois, Y. Hoffman & D. Pomarède, *The Laniakea supercluster of galaxies*, Nature **513**, 71 (2014)).

demonstrated in section 20, the strength of indirect interaction—that is, the interaction with outside masses—is expected to be negligible in comparison with the typical strength of a direct interaction in a local gravitational system.

The fact that the strength of the indirect gravitational interaction is negligible implies that possibility 1 has to be rejected. Thus, the only way to resolve the ratio (19) versus ratio (20) discrepancy is to suppose that possibility 2 is correct. That means that there is a local conflict caused by the execution of one or more other FLPs that results in the discrepancy.

Before I will examine possibility 2, let me present some additional arguments in favor of the negligibility of the influence of outside masses, which means the negligibility of indirect gravitational interactions. The negligible strength of indirect gravitational interactions is also implied by the high degree of accuracy of both Newton's law of gravitation and the theory of general relativity, neither of which accounts for the influence of outside masses. Einstein in particular showed that according to general relativity, the influence of outside masses is so "very feeble" that its "confirmation...by laboratory experiments is not to be thought of."[148] On the contrary, in his cosmological model Sciama advocated a strong influence of outside masses on a local gravitational system.[149] However, that influence was to be applicable only to the outside masses that are *very* distant. According to Sciama, the influence of "close-by" outside masses was negligible. He estimated that the contribution to "the inertia of the bodies on the earth, the whole of the Milky Way...[is] one ten-millionth, the sun one hundred-millionth and the earth itself one thousand-millionth," and that the remainder of the contributions are due to very distant masses.

As I will discuss in section 39, the gravitational influence of very distant masses on a local gravitational system is expected to be nil if the range of gravity is finite. The gravitational influence of not-so-distant masses on a local gravitational system is expected to be negligible according to the conclusion drawn in section 20. It follows that, from the perspective of the GPU theory, Mach's principle is not suitable to explain local gravitational effects.

The negligibility of the influence of outside masses on a local gravitational system can also be concluded from actual astronomic observations. To demonstrate this, let me refer to the lonely star – the galaxy-G scenario discussed previously in section 20, but let me now replace the lonely star with a BINARY-STAR system. From actual observations of binary stars, we know that the entire mass of galaxy G would have a negligible effect on the gravitational interaction between the two stars. The strength of the interactions between the stars in two identical binary star systems would be essentially the same regardless of the entire mass of the galaxy, except for the systems that are located very close to the bulge of the galaxy. The binary stars would also indirectly interact with the cluster of galaxies (let me call it "cluster C") that galaxy G is a part of as well as with the supercluster that cluster C is a part of. There seem to be no observations that would suggest a nonnegligible gravitational influence of galaxy G, cluster C, or the supercluster on the gravitational interaction between the binary stars.

25 The Force of Antigravity

Consider a galaxy in which the great majority of stars are located in its bulge, and assume that law III (the fundamental law of gravitation) is not being executed, whence the stars are not "pushed" toward each other. The stars are electrically neutral. Then, what could one expect to see as a result of the execution of the other FLPs? In particular, what motions of the stars would one expect to observe? I believe that the answer is rather simple if one carefully examines the FLPs identified in section 7. That examination suggests that the stars should be moving away from each other, pushed by a thermodynamic "force" generated as a result of the execution of the SLT. That is because nature eradicates nonequilibrium, which means that she forces disorder in the universe to increase. The mass of the stars in the galaxy is in a highly ordered state, as the majority of stars are grouped in the galaxy's bulge. A change toward a more disordered state means that the stars in the bulge should be moving outward—that is, away from one another.

Place two iron balls close to each other on a frictionless ice sheet far away from any other objects and assume, again, that law III is switched off. Consistent with the argument of the preceding paragraph, one would then expect the balls to move away from each other in response to the execution of the SLT. The cause of the motion would be a thermodynamic force. Let me call that force "the force of antigravity." It is reasonable to expect that the force of antigravity would act along the line connecting the gravity centers of the two balls. Now, let law III be switched back on. As a result, the force of gravity would drive the two balls toward each other—that is, toward a more ordered state. Then, one would be able to measure the *resultant* of (i.e., the magnitude of the conflict between) the true force of gravity and the force of antigravity. That resultant would be the apparent force of gravity (the magnitude of which could also be influenced by other conflicts with the execution of law III.) But one would not be able to measure the true force of gravity separate from the antigravity force, and vice versa. This is for the two forces are expected to act along the same line. There must exist one or more other conflicts with the execution of either law III or the SLT, or both, which make the magnitudes of the true force gravity and the force of antigravity different.

Also, there must exist a physical medium that carries the antigravity force. The gravitational field identified previously in section (19), which mediates the true force of gravity, seems to be a strong candidate for the medium that mediates the force of antigravity as well. That is because in both cases the forces are generated by interactions between masses. Note that the idea of a finite range of antigravity could not be supported. This is for an antigravity interaction between mass objects is expected to extend as far as the nonhomogeneity in the distribution of mass extends, as long as the masses are connected by a continuous gravitational field.

The execution of SLT is also expected to drive outward the matter that a mass object consists of—that is, it is expected to drive the matter of each mass object toward a more disordered state. On the surface, SLT loses the battle as the nuclear, electromagnetic and gravitational binding forces, which keep the matter of each mass object together, prevail. Yet, it appears to be a deceptive loss. I will submit later that the matter of a mass object actually is dispersed outward—that is, toward a more disordered state. This I will discuss in detail in section 31.

Therefore, it appears reasonable to conclude that it is primarily the antigravity force that reduces the strength of the true force of gravity to its apparent value. There simply is nothing else known that could result in such a reduction, which is identified by the comparison of ratios (19) and (20). Put otherwise, there seems to be nothing else known that would affect the execution of law III without affecting the execution of law IV in the same way. The existence of the force of antigravity allows for explaining the ratio (19) versus ratio (20) discrepancy.

The idea of an antigravity force of thermodynamic origin is, of course, radical and, to the best of my knowledge, new in physics. In this respect, I would like to hear an assertion that this idea is completely absurd followed, however, by an explanation of why the SLT affects ordered systems of energy but doesn't affect ordered systems of mass. One needs to keep in mind that according to the FLH, the SLT is an FLP and, as such, it is absolute. Hence, one cannot really justify excepting the applicability of the SLT to the eradication of mass nonequilibrium.

Implicit in the preceding discussion is the implication that an antiforce of thermodynamic origin does not exist in the case of two interacting charges. In this regard, one needs to recognize that the electrostatic force—generated as a result of the execution of law IV—does exactly the same as what a force of thermodynamic origin resulting from the execution of the SLT would be expected to do: "push" two charges toward each other, that is, push the charges toward equilibrium (toward a neutral state). Put otherwise, law IV can be considered to be a special case of the SLT, as suggested earlier in section 10. Hence, the execution of the SLT is not in conflict with the execution of law IV.

In summary, the fact that the executions of laws III and IV are expected to result in nonquantitative actions, which is a consequence of the simplicity-of-execution principle, allows one to conclude that there should be one or more conflicts with the executions of one or both of those laws that result in the ratio (19) versus ratio (20) discrepancy. Potential conflicts with other matter and energy can be eliminated based on the assertion that the outside (distant) masses can generate negligible gravitational interactions only in a local system (section 20). That, in turn, leads to the identification of the force of antigravity as the cause of the conflict with the execution of law III, which does not affect the execution of law IV.

In section 38, I will discuss the importance of the existence of the force of antigravity to the understanding of the structure of the universe.

26   Equivalence Principle

An implication of the simplicity-of-execution principle is that the contributions of two gravitationally interacting objects to the force of gravity acting between them are nonquantitative, which suggests that those contributions should be assumed to be equal for the purpose of formulating a physical law. That was the approach taken in section 22, in which a law of gravitation was derived. Such an approach, however, is applicable only where the amount of the mass of the object does not quantitatively reveal itself. An example of how the amount of mass may reveal itself is illustrated in figure 17. Let a ball falling from the Tower of Pisa collide with a ledge extending from the tower, whence a conflict between the execution of law III and other matter arises. It is only then that the ball's mass would reveal itself, as the amount of kinetic energy dissipated in the collision would increase with an increase in the mass of the ball. The amount of mass would also reveal itself after the ball comes to rest on the conflicting matter (on the ledge), since the strength of the contact force between the ledge and the ball would increase with an increase in the mass of the ball. For as long as the ball is in a free fall unobstructed by any conflicts with other matter or energy, or by a conflict with the execution of an FLP other than law III, the amount of

the ball's mass would not reveal itself in any way. (Recall from the discussion in section 10 that the quantitative response of matter/energy to the execution of an FLP is caused by conflicts with other matter or energy, or with the execution of another FLP.) Nonetheless, as pointed out in section (10), the execution of law III is in an inherent conflict with the simultaneous execution of law II. It follows that the mass of a ball could reveal itself even in a free fall.

Fig. 17 A falling ball gets into a conflict with other matter when it hits the ledge.

In the famous (apocryphal) experiment that established the principle of equivalence, two balls having different masses were dropped at the same time from the Tower of Pisa, as illustrated in figure 18. The initial velocities of the balls were both equal to zero. The observed acceleration

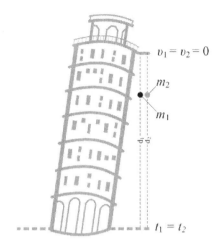

Fig. 18 Galileo's equivalence principle

of one ball was equal to that of the other ball at all times during the free fall, regardless of their masses, and both balls hit the ground at the same time. The Galileo's equivalence principle was thus established. As pointed out in the preceding paragraph, in the absence of conflicts with other matter or energy or with the execution of other FLPs, the simplicity-of-execution principle implies that the mass of a free-falling ball would not reveal itself. That means that the speeds at which the balls fall should not be affected by their masses and should be equal. It follows that if there were no conflicts of any kind with the execution of law III, Galileo's equivalence principle would become a consequence of the simplicity-of-execution principle.

The very foundation of general relativity refers to the well-known Einstein's happiest thought, "For an observer falling freely from the roof of a house, the gravitational field does not exist," which states the principle of equivalence. Based on the prior discussion, Einstein's happy thought can also be stated as: the mass of a free-falling observer does not reveal itself until she or he hits the concrete (hits an interfering matter). In the following discussion I will show that the equivalence principle is approximate only owing to a finite range of gravity—which, in turn, is a consequence of the FLH.

A ball falling from the Tower of Pisa is a result of the execution of law III (the law of gravitation). As explained in section 10, the motion of an object caused by gravitational interaction would be instantaneous if it were not for the conflict with the execution of the law of inertia. Because of that conflict, the amount of the mass of the falling ball would be expected to somehow reveal itself—that is, there should be a quantitative effect, the amount of which would depend on the mass of the falling ball.

An expedient way to appreciate that effect is to examine equation (16), which shows that the acceleration of a ball having mass m_2 would depend on its own mass (m_1 in that equation is assumed to be the mass of the earth, toward which the ball is falling). It follows that there would be a difference in the accelerations of the two balls having different masses and falling from the Tower of Pisa. Note that equation (16) includes a first-approximation-correction factor introduced in equation (14), designed to account for a finite range of gravity. Regardless of the mathematical form of that factor, however, its value has to depend on the masses m_1 and m_2 since both those masses determine distance d_L (defined in fig. 10), at which the acceleration of object m_1 must become zero. Consequently, the acceleration of an object depends on its own mass because of a finite range of gravity, from which the concept of distance d_L arises. Earlier in this section, I contended that the accelerations of the two balls falling from the Tower of Pisa can be expected to be independent of their masses. But, that contention was made for the scenario in which no conflicts with the execution of law III existed.

The connection between a finite range of gravity and the difference in the accelerations of two objects (m_1 and m_2) having different masses and falling toward object m_0 can be explained with the aid of figure 19. Objects m_1 and m_2 are located at the same distance from m_0. While the acceleration of object m_2 toward object m_0 is initially zero as the gravitational fields of those two objects are not connected, the acceleration of object m_1 has a greater-than-zero value as the gravitational fields of objects m_1 and m_0 are "overlapping." Let object m_2 be slightly moved toward object m_0 so that their gravitational fields just overlap. Object m_2 is now subject to a small, near-zero acceleration, while object m_1 is subject to a much greater acceleration since the overlap of the gravitational fields of objects m_1 and m_0 is substantial. It follows that the

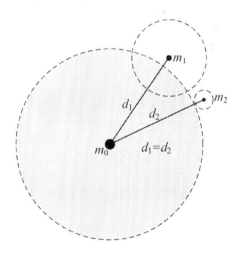

Fig. 19 The acceleration of object m_2 is zero or near zero while the acceleration of object m_1 has a definite value.

accelerations of m_1 and m_2, which are located at the same distance from object m_0, would be different. Thus, objects m_1 and m_2 would not collide with object m_0 at the same time. The conflicts that make the accelerations of the two objects different result from a finite range of gravity. One concludes, therefore, that the equivalence principle cannot hold exactly if the range of gravity is finite.

If the range of gravity is infinite, the accelerations of objects m_1 and m_2 would be the same for the objects that are located at the same distance from m_0. The correction factor introduced in equation (14) would have to be removed, and the acceleration of a falling object would be independent of its mass as the correction factor would be removed also from equation (16). That equation would become the same as the corresponding equation that results from Newton's law of gravitation, and the principle of equivalence would hold exactly. The mass of a falling object would not reveal itself in a quantitative way as the acceleration of the object would not depend on its mass. The effect of the conflicting law II (the law of inertia) on objects m_1 and m_2 would be the same at the same distance between those two objects and object m_0, regardless of the masses of the objects. Note that the difference in the accelerations of the falling objects is not directly caused by a finite range of gravity. It is caused by law II, the execution of which results in quantitatively different conflicts with the execution of law III (the law of gravitation): for a given distance, the amount of conflict caused by the execution of law II depends on the range of gravity of the falling object.

Note also that the difference in the times of the free falls of objects m_1 and m_2 illustrated in figure 19 is expected to be extremely small if the masses of the two objects are extremely small in comparison with the mass of the earth. (To appreciate, in quantitative terms, the extremely small difference in the free-fall times, see the results of t_L calculations in section 22.) This also means that the difference in the accelerations of the balls falling from the Tower of Pisa (fig. 18) would be extremely small even if the mass of one of the balls were equal to the mass of the Eiffel Tower, with the other ball having the mass of a matchstick.

The approximate character of the principle of equivalence is an unavoidable attribute of the universe that is infinite in time and space since such a universe implies the range of gravity to be finite (section 3). The exactness of the principle of equivalence has never been shown to be a carved-in-stone physical concept. It was just Galileo's conclusion drawn from experiments. Both Newton and Einstein (and others) built their theories of gravitation upon it. In this regard, it should be noted that many physicists actively pursue experimental evidence of that exactness. The Eöt-Wash group of physicists is one of the most active centers in that research and conducts, among other tests, a torsion-balance testing of the principle of equivalence. They recently reported a confirmation of the exactness of the weak equivalence principle to an accuracy of one part in 10^{13}.[150] In view of the foregoing discussion, that finding does not come as a surprise, as the test masses used in the Eöt-Wash experiments were extremely small (in the order of grams) in comparison with the masses of the attracting objects such as the earth, the sun, or the Milky Way galaxy.

27 Range of Gravitational Interactions

As pointed out previously, following Newton's publication of his law of gravitation, virtually all proposed cosmological models as well as the examinations of those models have been based on the oldest axiom of modern physics: the infinite range of gravity. In the GPU theory, that axiom is replaced with an implication of the infinite-time hypothesis, which leads to a finite range of gravity and the resulting postulate (1). Although that postulate is simple, it may be difficult to appreciate since the premise of an infinite range of gravity is deeply embedded in today's physics. To help appreciate the finite range of gravity, I will put postulate (1) into a quantitative perspective.

If the average mass density in gravitational field (D) is known, the range of potential gravitational interaction can be estimated for single-mass objects using equation (1). That range can be represented by the radius of the spherical gravitational field of a mass object. In a previous publication,[151] I estimated the value of D to be about 3.0×10^{-19} kg m^{-3} based on equation (14) and Asaph Hall's estimation of the power of distance d in Newton's gravity-force equation.[y] Hall determined that

---

[y] That value of constant D may seem very high. Yet the gravitational field would still be largely empty of matter and energy. At a mass density of 3.0×10^{-19} kg m^{-3}, a 1-kg mass would be contained, on average, in 3.3×10^{18} m^3 of space in a gravitational field. From another perspective, for stars with mass densities

Newton's equation, when applied to the condition of the mercury perihelion discrepancy, would work well if d^2 in that equation were replaced with $d^{2.00000016}$.[152] The above estimate of the average mass density originally appeared to be very attractive as it closely matched the average mass density estimated for the assumed volume of the Milky Way's gravitational field. I now believe that that estimate of the value D cannot be substantiated because calculating the same value for other planet and comet discrepancies yields different values of D. Nonetheless, the value of 3.0×10^{-19} kg m^{-3} represents a justifiable upper bound estimate of possible values of D because it is practically the same as the average mass density in the Milky Way's disk outlined by the *visible* stars and cosmic gas. The actual gravitational field of the Milky Way could be much larger than the size of the observed disk, whence the value of D could be much smaller. On the other hand, the results of the observations of how gravitational interactions affect the rotational speeds of stars suggest that the actual mass of the Milky Way could be ten times greater than the mass estimated by accounting for visible matter only.

The currently estimated value of mass density in the universe accounts for dark energy and dark matter. It is about 1.0×10^{-26} kg m^{-3} according to NASA and other sources.[153] I will use that estimate as a lower-bound value for the average mass density in the gravitational field simply because I have nothing else to go on.[z] Note that the value 3.0×10^{-19} kg m^{-3} is the average mass density in a gravitational field, while the value 1.0×10^{-26} kg m^{-3} is the average mass density in the universe. This presents a certain inconsistency in using those two estimates as the upper and the lower bounds, however, it will not meaningfully affect the following presentation.

The radii of the gravitational fields calculated based on the upper-bound estimate (3.0×10^{-19} kg m^{-3}) will be presented without brackets,

---

between 1×10^3 and 5×10^{17} kg/m^3, the fraction of space by stars would be between 3×10^{-22} and 6×10^{-37}, respectively.

[z] The 1.0×10^{-26} kg m^{-3} estimate of mass density in the universe corresponds to the CRITICAL MASS DENSITY, at which the UNIVERSE would be FLAT within the framework of the standard cosmological model. (The flatness of the universe is suggested from the sky surveys such as WMAP, BOSS, and others.) That means that the above estimate is based on the assumptions on which the standard cosmological model is based.

with the radii for the lower-bound estimate (1.0×10^{-26} kg m^{-3}) shown in brackets. While those two mass density estimates do not have any firm backgrounds, they seem to present a reasonable range of possible values. This is adequate for presenting the quantitative illustration of the finite ranges of gravity, which is the objective of this section. Even if those average density values were off by two or three orders of magnitude, the finite ranges of gravity would still be enormous, which is the point I want to make.

For a grain of sand with a 1-mm radius, the finite range of possible gravitational interactions is still comparatively gigantic and is in the order of 1.0×10^7 mm (3.2×10^9 mm). For a 1 kg steel ball with a radius of 3.1 cm, this range is about 9.3×10^7 cm (2.9×10^{10} cm). For the earth, which has a 6.4×10^3 km radius, it is about 1.7×10^{11} km (5.9×10^{13} km). For the sun, with a radius of 7.0×10^5 km, the range of gravity is about 1.2×10^{13} km (3.6×10^{15} km).

On the smallest scale, the ranges of gravitational interactions—as defined by the radii of gravitational fields—for gamma ray, visible light, microwave and ultra-low frequency photons would be in the order of 0.12 mm (37 mm); 0.0015 mm (0.47 mm); 0.000 046 mm (0.0143 mm); and 0.000 000 213 mm (0.000 066 mm), respectively.

(At the end of section 19, I suggested a simplistic model of a photon, in which the size of the photon is equal to the volume of the photon's gravitational field. In view of the radii of the gravitational fields of the various photons estimated in the preceding paragraph, it does not come as a surprise that ultra-low frequency photons can be used for communication with submarines, while gamma photons and other higher energy photons cannot penetrate water to any significant depth.)

In practical terms, the observations that could allow for a verification of Newton's assumption of the infinite range of gravity are limited to observations made in the solar-system. In the time of Newton, the farthest object confirmed to gravitationally interact with the sun was Saturn, which is situated about 9.0-10.1 AU from the sun. We know today that the distance to the edge of the visible universe is roughly 2.9×10^{15} AU. It follows that the premise of the infinite range of gravity introduced by Newton represented an extraordinary and most likely unprecedented-in-science extrapolation. (Mike Disney[154] is, and, I suspect, many other astronomers could be as baffled by the extent of that extrapolation as I am.) That extrapolation, the extent of which could not be appreciated at the time of Newton, is widely accepted today, while we know that the farthest known object that gravitationally interacts with the sun (the planet Eris) is situated about 96 AU from the sun, which still shows the extrapolation of the infinite range of gravity to be as extraordinary as the original Newton's extrapolation. Gravitational interactions have also been well examined in the nearby binary star-systems. But the separation distances of binary stars are still very small, typically less than 50 AU, so that a confirmation of the reasonability of Newton's extrapolation is not possible.

Finally, I have to say that I am not certain if it was really Newton who stated the axiom of the infinite range of gravity first. Perhaps, it was done later, by the physicists who interpreted Newton's law of gravitation. In this regard, Newton stated his thought in the *Principia*, "... the power of gravity ... operates ... according to the quantity of the solid matter which they contain, and propagates its virtue on all sides to immense distances, decreasing always in the duplicate proportion of the distances."[155] While the phrase "*always* in the duplicate proportion of the distances" could be interpreted to imply the infinite range of gravitational interactions, the word "immense" certainly does not mean "infinite."

CHAPTER IV: PHOTONS

If one wishes to appreciate the cosmology of the infinite-in-time universe in broad terms only, this chapter may be skipped if the following conclusion is taken for granted: the speed of the photon (the speed of light) must be constant and independent of the speed of the light-emitting source only when subject to gravitation. The key phrase of this conclusion is "only when subject to gravitation." To understand the reasoning underlying that conclusion and the mystery of the speed of light, one will have to read the entire chapter.

I want to emphasize that the ensuing explanation of the mystery of the speed of light (the speed of photons) is a suggestion made based on the SLT being applicable to the distribution of mass. Other explanations are, of course, possible. The fact is, however, that no other explanations appear to be available, and the constancy of the speed of light is simply assumed in today's physics to be a fundamental property of nature. (The concept of luminiferous aether, which was proposed in order to explain the physics underlying the constancy of the speed of light, has long been abandoned.) The problem with that assumption is that a photon could not then be a fundamental particle in the sense of the FLH, as it would have to have a speedometer and a speed-control mechanism built into it.

The proposed explanation of the constancy of the speed of light also is based on the FLH and the resulting simplicity-of-execution principle, as well as on a finite range of gravity, which is a consequence of the infinite-time hypothesis. As already stated, that explanation is based on the conclusion drawn from the FLH that the SLT is an FLP, which means that it is absolute and exact. This, in turn, means that the SLT forces the eradication of nonequilibrium in the distribution of mass.

28 Background

In this chapter, the unique characteristics of the speed of photons are discussed. They include the constant speed of the photon (any photon emitted from the surface of an object moves at a constant speed); the photon's speed being independent of the speed of the photon's emitting source; and the photon's speed being finite (equal to 3×10^8 m/s). Those three characteristics are together referred to as "the postulate of the constancy of the speed of light," which is one of the two key postulates underlying special relativity.[156] I believe that the credibility of any cosmological model would benefit from showing that the postulate of the constancy of the speed of light is consistent with the implications of the theory underlying the model. (In section 6, I suggested that the other key postulate underlying special relativity—the principle of relativity—can be viewed as a consequence of the FLH, which is a cornerstone of the GPU theory.) In this chapter, my intention is to suggest and explain the physical phenomena that underlie the constancy of the speed of light. In current physics, the constancy the speed of light is merely assumed based on both the implication of Maxwell's electromagnetic theory and the results of experiments.

The primary conclusion of the following discussion will be that the constant speed of the photon (light) results from the execution of the SLT and the existence of a gravitational field that represents a physical entity. The conclusion of key importance to the GPU cosmology will be that the speed of the photon has to be constant and independent of the speed of the emitting source only when subject to gravitation. This, in turn, will allow me to conclude that star formation out of thermal radiation is physically possible. That conclusion is of special importance since, as was pointed out in section 3, at least some stars have to be formed out of thermal radiation in the universe that is infinite in time. If the speed of the photon were constant regardless of the presence of a gravitational field, the premise of star formation out of radiation would be difficult, or perhaps even impossible, to substantiate.

29 Photon Energy

The energy (E) of a mass-energy object is given by the relativistic relation $E^2 = (pc)^2 + m^2c^4$, where p is the momentum, the term pc represents the kinetic energy of the object and c is the speed of light. Hence, the total energy of a massless ($m = 0$) photon is:

$$E_{ph} = pc. \qquad (21)$$

It follows that the total energy of a photon equals its kinetic energy, which depends on its momentum (motion) according to equation (21). The total energy of a photon can also be considered to be its intrinsic energy, which does not depend on motion. This is implied by the Planck-Einstein equation:

$$E_{ph} = hf, \qquad (22)$$

where h is Planck's constant, and f is the "frequency" of the photon. Equations (21) and (22) are related by de Broglie's relation between the momentum and the wavelength of a particle, $p = h/\lambda$, where λ denotes the wavelength, and the wavelength-frequency relation $\lambda = c/f$. The photon frequency incorporated in equation (22) depends on the surface temperature of the emitting body, while it does not depend on the photon's translational kinetic energy. It follows that the kinetic and intrinsic energies of the photon can be considered equivalent. This means that the total energy of the photon can be taken to be its kinetic energy or its intrinsic energy, depending on the context.

The equivalence of the photon's energies is valid only if the speed of the photon is constant and independent of the speed of an observer. If the speed of the photon were to depend on the speed of an observer, the photon's kinetic energy would depend on that speed, and the equivalence of its kinetic and intrinsic energies would not hold.

30 Maintaining the Speed of the Photon

Consider a region in the universe, denoted R_0, which is empty of matter and energy and contains zero Gravity—that is, it contains no gravitational field. Region R_0 is surrounded by a gravitational field. Any g-object, including a photon, located inside region R_0 is too far away from any other mass object in the universe to be subject to gravitational interaction. A Gravity-free region can exist because the range of gravity is finite. (In section 37, I will show that, in accordance with postulate (1), the existence of Gravity-free regions in the universe is necessary.) Let a photon moving with speed c enter region R_0 from the outside. Since a photon in the R_0 is not subject to gravitational interactions and is electrically neutral, it does not interact with any other energy or matter, which means that it is not subject to any known interactions.[aa] As a

---

[aa] Being electrically neutral, photons are unaffected by magnetic or electric fields.

result, the photon maintains its speed c consistent with the law of inertia. Let the photon collide with another photon and bounce off of it. The photon normally is assumed to continue to move with speed c after a collision. However, a collision could change the photon's speed, just like kicking a soccer ball changes the ball's speed. Since the photon does not interact with anything in the Gravity-free region R_0, it is on its own after the collision—that is, it is not affected in any way by any matter or energy, or any interaction. Therefore, to maintain its speed at c after the collision, the photon would have to measure its own positions, measure the times between those positions, calculate its speed, have a built-in mechanism (e.g., a jet engine) to control the speed, and then make decisions on how to use that built-in mechanism to maintain its speed at c. The idea that a photon could have such abilities and the idea of a built-in speed-control mechanism are so absurd that they have to be rejected. For all we know, a photon appears to be a fundamental particle in the sense of the FLH, which means that it doesn't have any measuring or decision-making abilities, mathematical skills, or built-in mechanisms. As a result, the premise that a photon, after a collision with another photon or another particle, must maintain its speed at c in a Gravity-free space cannot be upheld. In summary: where a photon is not subject to gravitation—that is, where no gravitational field exists—the photon's speed may be different than c.

Assume now that a region of the universe is still empty of mass and energy but filled with Gravity. Let a photon enter it with speed c and collide with another photon. After the collision, the photon will be subject to gravitational interaction, and it is known that it will move at the constant speed c, as this has been proven by the results of numerous measurements and experiments carried out in the earth's gravitational field—that is, in the space filled with Gravity. The foregoing discussion suggests that gravitational field is somehow responsible for maintaining the speed of the photon at c. That suggestion is a consequence of the proposition that a photon is a fundamental particle in the sense of the FLH and, as such, it does not make measurements, do mathematics, or have a built-in speed-control mechanism.

What could be baffling here is the question of how the speed of a photon is maintained. To explain this, let a massive particle be emitted from an object. The potential gravitational energy of the particle increases as it moves away from the object. The only energy available to compensate for that increase is the particle's kinetic energy equal to $1/2\, m \times v^2$, where v is the particle's speed. The intrinsic energy of the particle is not available to compensate for that increase. Thus, the kinetic energy of the massive particle must decrease owing to the energy conservation law. The decrease in kinetic energy results in a decrease in the particle's speed. The situation is different in the case of a photon. The speed of an emitted photon does not decrease, which can be explained by considering the energy of a photon. As pointed out in section 29, the intrinsic energy of a photon is the same as its kinetic energy. Consistent with the discussion of section 19, the energy-gravity equivalence implies that a photon is expected to behave like a massive particle in terms of gravitational interactions. Thus, a photon emitted from a mass object is expected to increase its gravitational potential energy. That increase occurs at the expense of the photon's kinetic energy, which is available in the form of the photon's intrinsic energy. It is that availability that, in terms of gravitational interactions, differentiates massless photons from massive particles. As a result, the speed of a photon can be maintained as its potential gravitational energy increases, while its intrinsic (or kinetic) energy decreases (the photon gets "cooler"). The decrease in the intrinsic energy of a photon emitted from an object results in a well-documented gravitational redshift. The increase in the intrinsic energy of a photon in motion toward a mass object results in a gravitational blueshift.

Region R_0, identified in the first paragraph of this section, presents a *true vacuum* in which nothing, not even Gravity, exists. (The concept of true vacuum was introduced in section 9.) If the range of gravitational interactions is assumed to be infinite, a true vacuum cannot exist. It is of interest to note that in special relativity, Einstein assumed the speed of the photon to be constant and independent of the state of motion of the emitting body in *empty space*, which he called "vacuo."[157] Yet he also believed in the infinite range of gravity, which means that no part of space could exist without a potential for gravitational interactions (or, in our terminology, without a gravitational field). Thus, Einstein did not recognize that the peculiar property of light (constant speed independent of the speed of the emitting source) could reasonably be expected to somehow relate to gravitation. Later, though, in general relativity, he did assume that light is subject to gravitational interactions. It also is worth noting that Einstein's view of the speed of light contradicted the view of Maxwell, who, in conjunction with his electromagnetic field theory, insisted that the propagation of light requires space to be filled with a medium necessary to carry light waves, which was later referred to as "luminiferous aether." In section 34, I will suggest that Gravity (the gravitational field) plays the role of Maxwell's aether.

31 Steady-State Speed of Radiation

The moon is hanging somewhere above the earth. The first thing we notice is that the moon is surrounded by a space that appears to be empty of matter and energy. As we perceive and compare that view of the moon with things on earth, we become fairly certain that the mass density in the space around the moon is much less than the mass density of the moon. (This has actually been confirmed by the astronauts who landed on the moon.) A high-school graduate who read sections 7 and 11 of this book would conclude that there is a definite nonequilibrium in the mass distribution between the moon and its surroundings. The graduate would instantly realize that the eradication of that nonequilibrium—that is, the dispersion of the moon's mass into its surroundings—should be happening because of the execution of the SLT. An undergraduate student of physics could relate the dispersion of the mass, which results in the eradication of mass nonequilibrium, to thermal radiation—that is, to the dispersion of mass in the form of photon energies. S/he could also make a rather appealing connection here and quote Einstein's famous statement, "If a body gives off the energy E in the form of radiation, its mass diminishes by E/c^2."[158] That student would also know that any mass object with a temperature above absolute zero emits radiation, which appears to be the most universal physical phenomenon (alongside the phenomenon of gravitation). There does not seem to be any known alternative to the conjecture that thermal radiation reflects the execution of the SLT when applied to the eradication of mass nonequilibrium—that is, to the eradication of *intrinsic-energy nonequilibrium*. This is because there doesn't seem to be another sufficiently universal physical process that would result in the dispersion of mass. "Universal" means here that the process is expected to occur regardless of the location of the photon's emitting body; its mass, dimensions, temperature, chemical composition; or its state of the motion. The dispersion of mass in the form of radiation energy is schematically illustrated in figure 20 for a single-mass object. It is the dispersion of mass that I want to discuss in this section so I can later arrive at an explanation of the constancy of the speed of the photon. As will be seen shortly, I could not explain it without concluding first that the range of gravity must be finite in the universe that is infinite in time (section 3).

Let the g-object shown in figure 6 comprise four electrically neutral objects. The four objects emit thermal radiation. Those objects and the emitted photons that are still within the gravitational field of the g-object, which has a volume of v_{1234}, do not directly interact with matter or energy located outside of that field, in accordance with the definition of a g-object (section 17). That is because those objects are neutral and, therefore, not subject to any other interactions with the outside matter or energy. From the perspective of mass contained in volume v_{1234}, the boundary of the gravitational field of g-object m_{1234} corresponds to a *zero-intrinsic-mass boundary* that delineates a thermodynamic system in nonequilibrium with respect to the distribution of mass. I will call that system TS_{1234}. TS_{1234} comprises the gravitational field of g-object m_{1234}, objects m_1 through m_4, and the emitted photons that are still within the gravitational field of object m_{1234}. The existence of a zero-mass boundary coincidental with the boundary of that field applies to any g-object.

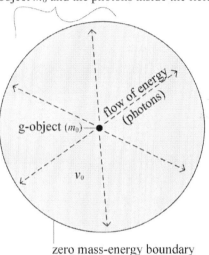

Fig. 20 Steady-state flow of energy in the execution of the SLT for a single-mass object.

Let the mass objects illustrated in figure 6 be ordinary objects such as stars or stones. TS_{1234} is in a state of thermodynamic nonequilibrium with respect to its mass distribution. That is because each of the four objects contain an enormous amount of mass in comparison with its surroundings. In other words, each of the four mass objects that form TS_{1234} is in a highly ordered (nonequilibrium) state. That means that those objects should be emitting their masses to bring the system toward a more disordered state, as required by the SLT. And they do, in the form of the emission of thermal radiation, as discussed earlier in this section.

The emmision and the motion of thermal radiation occurring as the result of the execution of the SLT, which drive a thermodynamic system toward equilibrium in mass distribution, does not seem to be recognized in normal physics.[bb] In this regard, I suggest the motion and the emission of thermal radiation to be the result of the execution of the SLT because I cannot find a sufficiently universal alternative to the dispersion of mass, which is necessary to bring an ordered mass system toward a more disordered state.

Figure 20 suggests that it is the gravitational field that mediates the dispersion of mass (the dispersion of intrinsic energy). That suggestion is based on the simple observation that there is nothing else that is known and present in the space containing the gravitational field of object m_0.

The execution of the SLT is illustrated in figure 21 for the heat energy nonequilibrium and the mass nonequilibrium. The difference between the two thermodynamic processes illustrated in that figure is the material (metal) versus the immaterial (gravitational field) media that carry the heat energy and the mass (in the form of radiation energy). As indicated in section 19, although the gravitational field comprises neither matter nor energy, it represents a physical entity as real as a metal plate. The flow of heat energy and the flow of radiation energy occur at steady-state speeds that are constant throughout the energy carrying media.

[bb] The generation of thermal radiation is a complex process. In simplest terms, it is a conversion of thermal energy into electromagnetic energy—the process that involves the thermal motions of charged particles which, in turn, release energy in the form of photons. On the quantum level, photons are released by atoms or ions in excited (elevated energy) states decaying randomly toward their lower energy states, that is, toward ground (equilibrium) states. Herein, I suggest that the generation of thermal energy is the result of the SLT driving a system toward the state of equilibrium in mass distribution. As will be seen from the discussion presented in the remainder of this section, the quantum view of thermal radiation (which is the generation of thermal energy) is not really essential to analyze the motion of the emitted photon.

The difference in the media that carry the radiation energy and the heat energy becomes essential in the more general scenario illustrated in figure 22. In that scenario, four single-mass objects emit mass (radiation energy), and four heat sources emit heat energy. With respect to the heat energy, the temperatures and the resulting heat flow gradients may vary across the metal plate. The variable heat (temperature) gradients result in variable speeds of the flow of heat energy across the plate. In the case of the flow of radiation energy, the situation is different. The gravitational field is the radiation-

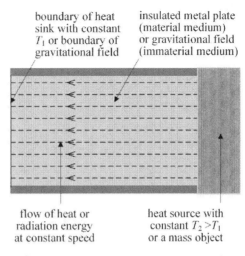

Fig. 21 Schematics of the steady-steady flow of heat energy or radiation energy.

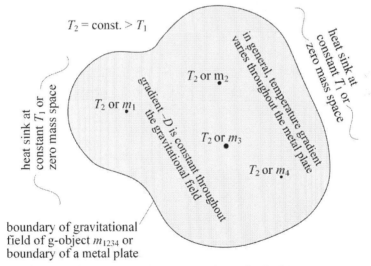

Fig. 22 Both the heat energy flow and the flow of radiation energy occur under a stead-state condition. In the case of heat flow mediated by the material medium comprising metal, the gradient will, in general, vary throughout the plate and so the speed of heat flow would vary. In the case of the flow of radiation energy through the immaterial medium comprising gravitational field, the gradient and the speed of energy flow are constant throughout the field.

energy-carrying medium, and it has only one property that is constant across the field: the average mass density in gravitational field (D). That leads to the constant gradient that drives the flow of radiation energy. The role of the average mass density in the emission of radiation energy is discussed in the remainder of this section.

The amount of mass nonequilibrium in a g-object, such as g-object m_{1234} illustrated in figure 6, can be determined by considering mass densities in the g-object's gravitational field. The mass density in the portion of the gravitational field of g-object m_{1234} outside of the four single-mass objects (D^*_{max}), which has the volume $v_{outside}$, is negligible as compared with the average mass density in the total gravitational field and can be assumed to be zero

$$D^*_{max} = 0. \tag{23}$$

The subscript "max" means that equation (23) describes the state of *maximum* mass nonequilibrium in TS_{1234}. The maximum nonequilibrium condition arises because the gravitational field outside of the four mass objects is essentially empty of mass. The condition of (hypothetical) *minimum* nonequilibrium, which corresponds to the state of perfect equilibrium, would occur if the total mass contained in the gravitational field of the g-object was uniformly distributed throughout the entire volume of the field. (Perfect equilibrium—i.e., the complete eradication of non-equilibrium—is the ultimate goal of any thermodynamic process enforced by the SLT.) That would correspond to a perfect-disorder condition within the system. According to postulate (1), the average mass density in the volume v_{1234} would then be

$$D^*_{min} = D. \tag{24}$$

It follows from equations (23) and (24) that the difference in mass nonequilibrium between the maximum and the minimum conditions concerning the distribution of mass—which defines the thermodynamic gradient that drives the system toward equilibrium—is

$$D^*_{max} - D^*_{min} = \frac{0-m}{v_{1234}} = -D. \tag{25}$$

where $m = m_1 + m_2 + m_3 + m_4$. According to postulate (1), the value of gradient $-D$ is constant and independent of the location in the gravitational field and time. Thus, the flow of radiation energy always occurs at the same steady-state condition as it is driven by a constant gradient. It follows that the speed of the flow of radiation energy (the speed of the photon) is the same at any location in the gravitational field, consistent with any other steady-state thermodynamic process. The steady-state flow of radiation is possible because the boundary of the thermodynamic system is open with respect to the flow of energy and because, from the perspective of the mass-energy contained inside the gravitational field, that boundary represents a perfect sink with respect to the outflow of radiation energy (mass).

Thus, the average mass density in the gravitational field plays the role of a gradient in the thermodynamic process of the flow of the energy emitted by a mass object. The same can be concluded from a more familiar consideration. Consider a single-mass object such as that shown in figure 20. There is mass m_0 at the center of the gravitational field and mass equal to zero along the boundary of that field, with the difference, $0 - m_0$ spread over the volume of the gravitational field (v_0). Thus, the mass gradient is $(0 - m_0)/v_0 = \nabla m = -D$. The gradient is expressed in units ML^{-3}. This dimension is consistent with the concept of gravitational field, which is quantized by its *volumetric* value v (see sections 17 and 19). The gradient $-D$ is rather analogical to the temperature gradient: there is a constant temperature (T_1) at the left side of the metal plate illustrated in figure 21 and a constant temperature (T_2) at the right side of the plate, with the difference spread over the length of the plate (L), which define the heat flow gradient as $(T_1 - T_2)/L = \nabla T$. Under those conditions, the temperature gradient is the same at each location of the metal plate, and the heat energy always flows at a constant speed.

In summary, in respect of the gravitational field of any g-object, the average mass density (D) represents the thermal radiation flow gradient which, in accordance with postulate (1), always has the same magnitude regardless of time and location in space, its motion, mass, and chemical composition. It enforces a steady state radiation flow process.[cc]

[cc] Assume that object m_0 shown in figure 20 is so massive (and cold) that the entire intrinsic energy of a photon emitted from that object is converted into gravitational potential energy before it reaches the boundary of the object's

32 The Speed of the Photon

I will first examine the speed of the photon emitted by a single-mass object that is a constituent of a g-object such as, for instance, g-object m_{1234} (figure 6). As discussed in the preceding section, the motion of the emitted radiation energy occurs under a steady-state speed, which is independent of the direction of motion, the location (x_i), or the time (t). Thus, the speed of the photon (v_{ph}), which is enforced by the constant thermodynamic gradient $-D$, is the same for all emitted photons. The speeds of the photons are measured in the frame of reference of the g-object's gravitational field. Therefore, in that frame of reference,

$$v_{\text{ph}} = \text{const.}(\vec{x}_i, x_i, t) = c. \tag{26}$$

where $\vec{x}_i$ is a unit vector in the direction of the photon's motion. Consistent with the common convention, the speed of the photon in equation (26) is denoted as c. (Equation such as $A = \text{const.}(x, y)$ states that the value of A does not depend on the values of x and y.)

The above conclusion concerning the constant speed of the photon requires an explanation of the role of the surface temperature of the photon-emitting object. To appreciate that role, consider two mass objects that are identical in all respects except that their surface temperatures are different. According to the Stefan-Boltzmann law

$$\frac{\Delta E}{\Delta t} = \sigma A T^4, \tag{27}$$

where σ is a constant, A is the surface area of the radiation-emitting object and T is the surface temperature of the object. It follows that the

---

gravitational field. (That conversion was discussed in section 30.) The photon ceases to exist. The gravitational field of the object shrinks, extremely slowly, as the intrinsic energy of photons is converted into gravitational potential energy. The system becomes a "black hole" as no electromagnetic radiation escapes it. However, very small amounts of gravitational potential energy are emanating from the system. Whether so extremely massive objects can or do exist—I do not know. I want to emphasize that the black hole described here is different from the Schwarzschild's black hole, which I consider to be a mathematical artifact (being a singularity found in a solution of Einstein field equations).

rate of the energy flow ($\Delta E/\Delta t$) would be different for the two objects. However, the *speed* of the energy flow—that is, the speed of the photon—would be the same for both objects regardless of their surface temperatures. That is because the thermodynamic gradient $-D$, which drives the speed of the photon, does not depend on temperature. Thus, the surface temperature of the emitting body does not influence the speed of the emitted photons. That is much like a speed limit set for a highway, with all motorists obediently observing that limit: during a rush hour, the rate of the car flow would be higher, but the speed of the cars would be the same as it would be during the small hours of the night. Nonetheless, it needs to be kept in mind that the distribution of heat energy in the system comprising a photon-emitting object and its surroundings also is subject to the SLT eradicating a thermal nonequilibrium. In this process, however, $-D$ is not the gradient that drives the speed of eradication.

So far I have limited the discussion of the speed of the photon to the photons emitted in the gravitational field of a g-object and have neglected the photons that may enter the gravitational field of a g-object from outside. Those photons would also move at speed c. Their speeds would be enforced by the same steady-state thermodynamic process, under the same gradient ($-D$). That is, in executing the SLT, nature does not differentiate between a photon emitted in a local gravitational field (e.g., in the solar system) and a photon emitted in an "overlying" gravitational field (e.g., a photon emitted from another star in the Milky Way that passes through the gravitational field of the solar system). Put otherwise, while the premise of constant thermodynamic gradient was concluded from the consideration of g-object m_{1234} shown in figure 6, this conclusion would also hold if g-object m_{1234} were replaced with the solar system or the Milky Way galaxy.

The foregoing conclusion can be generalized as follows: any photon emitted from a mass object will move with constant speed c in relation to the gravitational field as long as it remains within the field—that is, as long as it is subject to gravitation. This is consistent with the conclusion drawn in section 30: "gravitational field is somehow responsible for maintaining the speed of the photon at c.". On the contrary, in a Gravity-free space, there is no thermodynamic gradient $-D$. One concludes, therefore, that a photon can move at any speed when it is not subject to gravitation.

The above conclusion is analogous to the conclusion that can be drawn with respect to the flow of heat energy. The speed of the heat energy flow is constant as long as that flow is controlled by a constant heat gradient. This is illustrated in figure 21. The speed of heat energy flow is constant as the energy flows through the metal plate under a constant gradient. After the heat energy exits the plate, its speed of motion can be different as the flow of energy is no longer controlled by that gradient. If the metal plate is surrounded by air, the speed of the heat energy flow outside the plate would depend on the gradients decided by the temperatures and the motions of the surrounding air molecules.

Figure 23 illustrates the gravitational field of a galaxy, in which a moving star and two observers (Obs. 1 and Obs. 2) are located. The observers are stationary in relation to each other. The star and the star's gravitational field move with speed v_s relative to the observers. Two photons (Ph. 1 and Ph. 2) are emitted from the surface of the star toward each of the observers. Consistent with the prior discussions, both photons move with speed c in relation to the star's gravitational field as well as to the galaxy's gravitational field.[dd] By definition, all stars and other single-mass objects are stationary in relation to their individual gravitational fields, which, in turn, form the gravitational fields of the overlying g-objects such as galaxies or galaxy clusters. It follows that single-mass

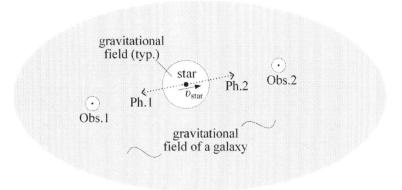

Fig. 23 The speed of the photon does not depend on the speed of the photon's emitting star. Both observers and the star are stationary in relation to their gravitational fields.

[dd] The gravitational field of a galaxy must be continuous, that is, there cannot be Gravity-free regions embedded in it.

objects are also stationary in relation to the gravitational fields of the overlying g-objects. Thus, the two observers will measure the photon speeds to be the same regardless of the speed of the moving star, which also is stationary in relation to the gravitational field of the galaxy. (The star is moving relative to the observers but not relative to the surrounding gravitational field.) That means that photons "carried" by the gravitational field will pass both observers with speed c in the frame of reference of the galaxy's gravitational field, which is stationary in the frame of reference of the observers. The mystery of the speed of the photon being independent of the speed of the photon's emitting source is thus explained: the speed of the photon in motion through gravitational field does not depend on the speed of the photon's emitting object because the speed of the photon is the same in any gravitational field and because all objects, including all observers, are stationary with respect to that field.

For a nonspecialist, the foregoing explanation of the speed of the photon being independent of the speed of the emitting source could be difficult to grasp. A simpler, while less formal explanation is: recall from the discussion presented in section 31 that the thermodynamic gradient $-D$ does not depend on the speed of the photon's emitting source and is constant throughout any gravitational field. Specifically, it followed from that discussion that the thermodynamic gradient did not depend on the individual speeds of the four single-mass objects illustrated in figure 6. Therefore, the thermodynamic gradient would not depend on the speed of the star shown in figure 23. Since that gradient determines the speed of the emitted photon, the speed of the photon does not depend on the speed of the photon's emitting source (i.e., on the speed of the star).

What still can be baffling here is: how is it possible that the moving star is stationary in relation to the galaxy's gravitational field? Look at figure 23 again. By definition, the star is stationary in relation to its own gravitational field (the white circle that surrounds the star). But, one might argue, the white circle is moving in relation to the gravitational field of the galaxy, so the star must be moving as well. Wrong argument. It could only be invented by the human mind. Nature in executing her laws, including the SLT, is unable to use the motion of the star to generate a quantitative effect. She cannot know that the star is moving. Neither she knows that the photons were emitted from a star. Remember, she has no memory. She is unable to identify the white circle and check whether the emitted photons are still within or outside of it. As far as she is concerned, there is only *one gravitational field with exactly the same property throughout*. It is the gravitational field of an overlying g-object (e.g., a galaxy cluster or a supercluster). The white circle has no meaning in respect of the speed of the photon. It cannot be distinguished from the remainder of the gravitational field. It is only the gradient that counts, and that gradient is exactly the same inside and outside of the white circle. The speed of the photon in the frame of reference of the overlying gravitational field will be the same regardless of the speed of the star. One concludes, therefore, that in any gravitational field

$$v_{\text{ph}} = c = \text{const.}(\vec{x}_i, x_i, t, v_{\text{source}}), \tag{28}$$

where v_{source} denotes the speed of the photon's emitting source. The speed of the photon is independent of the properties of the emitting source, such as the source's mass, dimensions, temperature, chemical composition, as well as its speed or orientation,

In summary, while objects (sources of radiation and observers) can be moving in relation to one another, they are always stationary with respect to the surrounding gravitational field. As a consequence, the speed of light will always be observed to be constant and the same for any observer regardless of the speed of the observer or, equivalently, the speed of the emitting source. This is because the speed of the photon always has to be the same in relation to any gravitational field. The Maxwell's 150-year-old question, "In relation to what frame of reference is the speed of light constant?" can now be answered: "in relation to the gravitational field."

In light of the above discussion, it does not come as a surprise that the famous Michelson-Morley experiment failed to reveal aether.[159] As illustrated in figure 24, the speed of the light beam is expected to be the same regardless of the direction and the speed of the earth's motion in relation to other objects. That is because in the earth's gravitational field light beams travel at speed c in the frame of reference of that field, which is stationary in the earth's frame of reference. I note that the idea of an aether that is stationary in relation to matter is old. It was put forward by George Gabriel Stokes in 1840s (see, e.g., Michel Janssen and John

Stachel[160]). Stokes's idea of stationary aether drag was proposed in order to replace the concept of "entrained" aether that was introduced by Augustin-Jean Fresnel in 1818. That aether allowed for the relative motion of matter and aether. It is of interest to notice that the existence of any kind of aether was rejected in 1905 by Einstein in the theory of special relativity.[161]

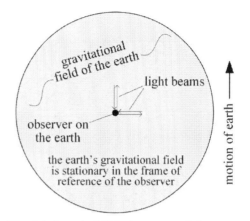

Fig. 24 Speed of light is measured the same regardless of the earth's motion.

As pointed out in section 28, there are three basic characteristics of the speed of light. The first two, which include the constancy of the speed of the photon and the independence of that speed from the speed of the photon's emitting source, have now been explained in physical terms. The third characteristic, which is the finite speed of the photon, is discussed in the next section.

It seems to me that the most persuasive argument in favor of the constant speed of light being the result of a thermodynamic process, could be the outcome of the experiment that has been carried out thousands of times. Let a light beam enter and exit a body of glass. The speed of the light beam upon entering and after exiting the body would be exactly the same and equal to c, while that speed would be lower when the beam runs through the glass. Now, why would those entering and exiting speeds be the same? If the light beam—that is, each photon of which the beam consists—carries a speed measuring device and a jet engine to control its speed, then the answer is simple: each photon knows that it has to move outside of the glass with speed c, so it measures its own speed and then fires the engine to return to speed c upon exiting the glass. Still, that answer is even simpler if the photons are fundamental particles—that is, they do not carry any measuring devices or jet engines, they do not know the value of speed c, and do not make any decisions. Upon exiting the body of glass, the motion of the photon is subject to the same thermodynamic gradient that controls its speed as the gradient existing prior to entering the body of glass. When moving through the glass, the motion of the photon is subject to a relatively strong conflict between the execution of the SLT and the interfering matter (glass), which results in a reduction of the speed of the photon. In the space that is empty of matter and energy, there would be no conflict between the execution of the SLT and other matter or energy, and the speed of the photons would be equal to c.

33 The Finite Speed of the Photon

The speed of the photon emitted as a result of the execution of the SLT should be infinite according to the statement of the SLT (law I in section 7). Yet we know that the speed of the photon is finite. Therefore, there must exist one or more conflicts with the execution of the SLT. (In this regard, recall the conflict with the execution of the SLT in the case of the metal rod illustrated in fig. 1: the matter that the rod is made of and the interfering FLPs result in conflicts that prevent the thermal nonequilibrium from being eradicated instantly.)

Consider an emitted photon that is moving in a space that is free of matter and energy. The space, however, is filled with gravitational field. The photon's speed is expected to be limited to a finite value by a conflict that is mediated by the gravitational field. That is because the photon is electrically neutral, and there is nothing else in the space free of matter and energy. Possible conflicts include either interactions with outside masses or local conflicts with other FLPs. These correspond to the two possibilities discussed in section 24. Again, the influence of outside masses on the motion of the photon is expected to be negligible, whence the only possibility left is that conflicts with other FLPs limit the speed of light. Candidates for the conflicting FLPs include laws II, III, and VI identified in section 7. For instance, law II (mass resists a change to its state of motion) would be expected to prevent the instantaneous motion of the photon. Law VI (the momentum-conservation law) would be expected to prevent the emitted photon from moving with infinite speed since, if it did move with infinite speed, the emitting body would have to recoil with infinite speed as well to satisfy the momentum-conservation law. That, of course, is unconceivable.

The idea that photons might interact with a gravitational field is old. For instance, in 1937 Hubble wrote, "Internebular space, we believe, cannot be entirely empty. There must be a gravitational field through which the light-quanta travel for many millions of years before they reach the observer, and there may be some interaction between the quanta and the surrounding medium. The problem invites speculation, and, indeed, has been carefully examined."[162] While the context of Hubble's speculation was different (Hubble referred to the tired-light hypothesis), his idea of the interaction between a gravitational field and photons is somewhat similar to the idea of conflicts with the motion of the photon mediated by the gravitational field.

In section 18 I argued that the speed of gravity can be infinite. That is consistent with the statement of law III (section 7). In this chapter, I have reasoned that the speed of the photon has to be finite when subject to gravitation. That reasoning was based on the submission that the SLT executes the eradication of nonequilibrium in the distribution of mass. However, similar reasoning cannot be applied to the speed of gravity, or the speed of any other interaction that involves neither matter nor energy. Because such an interaction in itself comprises neither matter nor energy, it cannot be affected by any matter or energy. This, of course, is contrary to the motion of the photon (light), which *is* the motion of energy.

34 Maxwell's Aether and Einstein's Vacuum

In conjunction with his electromagnetic theory, Maxwell insisted that a physical aether that permeates space and mediates electromagnetic waves should exist so that his conclusion of the constancy of the speed of light would be supported. Einstein, however, argued in conjunction with presenting the theory of special relativity that the existence of aether is superfluous and assumed the speed of light *in vacuum* to be constant (as there is nothing to interact with in a vacuum, he effectively assumed that the photon has a speed measuring device and another device that can be used to somehow maintain the same speed in relation to any observer).

It seems that the problem with Maxwell's aether versus Einstein's vacuum can be resolved if the gravitational field introduced in section 19 is recognized to be a physical entity that plays the role of Maxwell's aether. As discussed in this chapter, the gravitational field mediates the motion of electromagnetic radiation, which results in a constant speed of the photon in relation to any observer. And, as suggested in section 19, the gravitational field (Gravity) has an undeniable physical meaning.

CHAPTER V: UNIVERSE

I would like to start this chapter, which is intended to explain how the infinite-in-time universe works, with the listing of the conclusions that were drawn in the prior, background chapters. They primarily were drawn from the consideration of the fundamental laws of nature and logical reasoning.

35 Summary of Conclusions Drawn in Chapters I through IV

From the perfect cosmological principle assumed to hold in section 2 and the resulting infinite-time hypothesis, the following conclusions were drawn:

- The universe is infinite in time.
- The universe is spatially infinite.
- The universe is nonexpanding and nonevolving on a large scale.
- The entropy in the universe is constant.
- The number of galaxies in an RVU is expected to be much higher than is currently estimated.
- CMB is expected to exist with very low-temperature (invisible) galaxies as its source.
- At least some stars have to be born out of radiation and die by emitting their entire energies.
- Therefore, stars emit more energy than they absorb.
- The range of gravity is finite.

From the fundamental-law hypothesis and the fundamental laws of nature identified in section 7, the following conclusions were drawn:

- The fundamental laws of nature (physics) are extremely simple and nonmathematical, and exist independent of the human mind.
- The typical physical laws that describe physical phenomena in quantitative terms (i.e., the "mathematical models" or "predictive theories") are invented by physicists and exist in the human mind only.

- The physical laws reflect the execution of the fundamental laws of nature and conflicts with the execution of other fundamental laws and/or conflicts with interfering matter or energy.
- Physical space and physical time come into existence as a result of those conflicts.
- The FLPs are simple in the extreme. The executions of the FLPs are nonquantitative.
- The action-at-a-distance mode of physical interaction is rational in gravitational field, but not in a true vacuum.
- Mathematics appears to be unreasonably effective in advancing applied physics because of the similarities between mathematics and physics.
- The fundamental problem with physics appears to be the current paradigm of physics, which humanizes nature.

From postulate (1), which presents the relation between a finite range of gravity and mass, the following conclusions were drawn:

- Gravity (gravitational field) is a physical entity.
- The range of gravity and mass are directly proportional.
- Owing to a finite range of gravity, Gravity-free regions can exist in the universe.
- Gravity and energy are equivalent in the sense that one cannot exist without the other.
- Gravity is conserved, which states the gravity-conservation law.
- The speed of gravity can be infinite as there appears to be no known physical phenomenon that could put a limit on the speed of physical interaction that involves neither matter nor energy.
- The gravitational influence of distant (outside) masses on a local gravitational system is expected to be negligible.
- Newton's law of gravitation can be theoretically derived from the consideration of the simplicity-of-execution principle and postulate (1). The derivation involves the averaging of basic variables over space and over a period of time.
- The strength of the force of gravity is in reverse proportion to the average mass density in the gravitational field.

- The ratio of electrostatic force to gravity force, which is huge as the interactions between fundamental particles are considered, can be explained by the action of a thermodynamic force named "the force of antigravity."
- The force of inertia cannot be explained by Mach's principle.
- The equivalence principle, while highly accurate, is approximate only due to a finite range of gravity.
- The range of gravitational interaction of any single mass object, while finite, is still enormous in comparison with the size of the object.

From the premise of the finite range of gravity and from the submission that the emission of thermal radiation results from the execution of the SLT, the following conclusions were drawn:

- The speed of a photon not subject to gravitational interaction may be different from c.
- Thermal radiation is emitted in a steady-state process that is controlled by a constant thermodynamic gradient.
- The speed of a photon subject to gravitational interaction is constant and independent of its location in the gravitational field, the time, and the speed of the emitting source.
- The speed of an emitted photon can be constant because its intrinsic energy is available to compensate for the change in the photon's gravitational potential energy.
- The finite speed of a photon, driven by the execution of the SLT, can be explained by the photon's interactions with other FLPs, which are mediated by the gravitational field.

36 Background

In today's cosmology, astronomical observations are interpreted according to the assumptions and the implications of the standard cosmological model, and no other interpretations are normally pursued. In this chapter, I will discuss the primary features of the cosmological model of a nonevolving and spatially-infinite universe, and show that that model allows for explaining many, if not all, of those observations.

A most notable feature of the GPU model is that all its suppositions are experimentally verifiable, at least in principle.

A brief outline of the GPU cosmology was presented in section 3. The discussion in this chapter focuses on the filamentary structure of the universe comprising the GALAXY FILAMENT and the cosmic voids (the "cosmic web"), which was discovered from the redshift surveys in the late 1970s and early 1980s by M. Joeveer et al.,[163] S. A. Gregory and L. A. Thompson,[164] R. P. Kirshner et al.,[165] J. Huchra et al.,[166] and others, as summarized by J. R. Bond et al.[167] This was one of the most important discoveries in modern astrophysics, and investigations of the cosmic web have dominated astrophysical research ever since. A comprehensive summary of that research was presented by Rien Van de Weygaert and Erwin Platen.[168] In the standard cosmological model, cosmic voids are assumed to originate from local minima in the primordial density field and to be further developed by gravity (see, e.g., E.G.P. Bos et al.[169]). Consistent with the standard model, cosmic voids must have been expanding faster than the galaxy filament in order to acquire their typical size (see, e.g., Jounghun Lee and Daeseong Park[170]). Jeremy Tinker and Charlie Conroy showed that such a development of cosmic voids can be described using the standard model.[171] Physically, empty space in a cosmic void is no different from empty space in a galaxy filament, according to the standard cosmological model. In the GPU model, those two spaces are expected to be fundamentally different in physical sense.

Another key cosmological feature of the universe discussed in this chapter will be the birth and the death of a typical star. It will be argued that the speed of a photon, which can be less than c when not subject to gravitation, makes the birth of a star out of the energy of electromagnetic radiation possible. This is the radiation that originates from all stars and from gas and dust in the universe. Finally, new explanations of key astronomic observations, such as the observed distribution of galaxies, cosmological redshifts, and the CMB, will be presented.

37 The Insufficient-Mass Conclusion

Consider a scenario in which a representative volume of the universe (RVU) contains an amount of mass such that it is fully filled with Gravity (with gravitational field). In accordance with postulate (1), the amount of mass contained in the RVU would then be equal to

$m_{RVU} = RVU \times D$—that is, the average mass density in the universe would be $D_{RVU} = D$. Under that scenario, it would be impossible to add more mass to the RVU, since this would violate postulate (1). However, the removal of some mass from the RVU would violate neither postulate (1) nor any of the conclusions drawn in the preceding discussions. Then, regions empty of gravitational field, which I will call "Gravity-free regions," would exist in the universe, and the average mass density in the universe would be

$$D_{RVU} < D. \tag{29}$$

The existence or nonexistence of Gravity-free regions in the universe depends on the amount of mass in an RVU. There is only one (maximum possible) amount of mass, equal to m_{RVU}, that would result in a complete filling of the RVU with Gravity. On the other hand, there are an infinite number of smaller mass amounts that would result in a partial filling of the RVU with Gravity. Therefore, the probability that the actual amount of mass in the RVU is less than m_{RVU} is infinitely larger than the probability of that amount being equal to m_{RVU}. From this consideration, it is reasonable to suppose that the amount of mass in an RVU is less than the mass that would be required to fully fill the RVU with Gravity. I call this supposition the *insufficient-mass conclusion*. As I will discuss in later this chapter, that conclusion allows for a rational explanation of some key astronomical observations, which presents an argument in its favor.[ee]

The insufficient-mass conclusion can also be seen as a consequence of the perfect cosmological principle, from which it was concluded that the range of gravity was finite (Section 3). To explain that inference, let me first observe that the sequence of gravitational units in the universe must be finite if the cosmological principle holds. The basic gravitational unit is a single-mass object such as a star. A gravitational unit at the next

---

[ee] From a philosophical perspective, the assumption that the amount of mass in an RVU is exactly equal to m_{RVU} would be unsatisfactory. Such an assumption could be perceived to imply the existence of a designer-builder (a creator) who chose the specific amount of mass and then supplied that amount. Yet in the universe that is infinite in time, there is no place for a creator or an act of design-creation.

level would be a g-object such as the solar system or a binary-star system (assuming that the constituents of the solar system or the binary star system do not directly interact with other cosmic objects). A galaxy would represent a gravitational unit at the next level, followed by a cluster of galaxies and then by a supercluster, which is currently believed to be the largest gravitational unit in the universe. Owing to the cosmological principle, the sequence of gravitational units cannot be infinite because, under this circumstance, that principle would not hold. (An RVU would have to be infinitely large for that principle to hold.) So let me suppose, consistent with the suggestion emerging from astronomic observations, that superclusters are indeed the largest gravitational units. This means that superclusters are the largest g-objects in the universe. Even if larger gravitational units did exist, the following arguments would still be valid.

Consider a scenario in which there are no Gravity-free regions in the universe—that is, $D_{RVU} = D$. In that scenario, each segment of the boundary of the gravitational field of a supercluster would have to coincide exactly with a segment of the boundary of the gravitational field of an adjacent supercluster. That means that wherever such a boundary is deformed due to the motion of masses inside a supercluster (e.g., the motion of a galaxy or a cluster of galaxies), the exact "matching" and opposite motion of masses would have to take place in the adjacent supercluster. Otherwise, the two gravitational-field boundaries would separate or overlap, which cannot happen as it would violate postulate (1). Such a correlation between the motions of masses in adjacent gravitational fields would require the existence of a law of nature that monitors and controls those motions. That would require nature to measure or predict the motions of masses and then to apply some means to make the motions in adjacent superclusters opposite while exactly matching. The existence of such a law of nature would violate the FLH, whence the notion of nature's monitoring and controlling the motions of an infinite number of mass objects has to be rejected. Therefore, D_{RVU} cannot be equal to D and its value has to be less than the value of D. That leads to the conclusion that Gravity-free regions in the universe have to exist. The boundaries surrounding Gravity-free regions, which also represent the boundaries of the gravitational fields of superclusters, can then deform in response to the motions of masses inside superclusters, with no overlapping of their gravitational fields.

To better appreciate the above conclusion, take a pencil and a piece of paper and draw a near-vertical wavy line. This line is meant to portray a joint boundary between the gravitational fields of two (right and left) superclusters. Now on both sides of the line, draw some spots intended to represent galaxies. In your mind, move the spots on the right-hand side of the joint boundary to the right. The boundary of the gravitational field of the right supercluster will move to the right, and a gap will be formed between the gravitational fields of the superclusters. This will happen unless the galaxies on the left side of the wavy line move in exactly "matching" directions such that the boundaries of the gravitational fields of the superclusters do not separate. For the galaxies to move in exactly "matching" directions, nature would have to intervene by executing an "exact match" law, which cannot exist according to the FLH. Note that owing to postulate (1), a gap cannot form if $D_{RVU} = D$. Therefore, this assumption has to be incorrect. A gap can form if $D_{RVU} < D$. In this case, Gravity-free regions have to exist in the universe and the deformation of a supercluster boundary at one location can be compensated for by extending that boundary at another location, into a Gravity-free region. Thus, the total volume of the supercluster's gravitational field can remain unchanged, as required by the mass conservation law and postulate (1). If the masses on both sides of a superclusters' joint boundary move toward each other, the total volume the two gravitational can remain unchanged, as before, by extending their boundaries into one or more Gravity-free zones. In both cases, the amounts of Gravity and Gravity-free volumes in an RVU remain the same. To summarize, if postulate (1) holds, Gravity-free regions in the universe must exist so that galaxies can be in motion.

38 The Structure of the Universe

In the GPU cosmological model, the gravitational structure of the universe is derived from the cosmological principle, postulate (1), and the insufficient-mass conclusion. Those imply two possible scenarios that could describe the gravitational structure of the universe. In the first scenario, the structure of the universe comprises a web-like gravitational field, which extends to infinity and contains bubble-like Gravity-free regions that are distributed homogeneously on a large scale throughout the universe, consistent with the cosmological principle. The average mass density in the gravitational field is D, as required by postulate (1), and the average mass density in the universe (D_{RVU}) is less than D, as required by inequality (29).

In the second scenario, the gravitational structure of the universe comprises a web-like Gravity-free region that extends to infinity and contains bubble-like gravitational fields distributed uniformly throughout the Gravity-free web. The average mass density in each gravitational bubble is equal to D, and the average mass density in the universe is less than D, as required by inequality (29). The problem with this scenario is that given infinite time, the masses in each of the gravitational bubbles (in each gravitational field) would have already collapsed to its gravity center, which, of course, is not the case. Consequently, one concludes that it is the former scenario that properly identifies the gravitational large-scale structure of the universe. That scenario is schematically illustrated in figure 25.

The web-like gravitational field and the Gravity-free regions are identified with the galaxy filament and some or all of the cosmic voids, respectively—that is, with the structure of the universe discovered from the redshift surveys (section 36). The galaxy filament and the cosmic voids are delineated by Gravity – no-Gravity boundaries, as illustrated in figure 25. The gravitational fields of stars and galaxies may exist inside cosmic voids. Those fields would be gravitationally separated from the gravitational field that forms the galaxy filament. Observations indicate that cosmic voids are regions of the universe that are largely empty of matter—that is, they contain very few or no galaxies. Thus, the existence of Gravity-free cosmic voids is consistent with those observations.

The average effective diameter of a cosmic void is about 30 Mpc, or 1×10^8 ly (e.g., Foster and Nelson[172], Pan et al.[173]). The diameter of the gravitational field of a typical galaxy with a mass of two hundred billion solar masses is between 7×10^3 and 3.5×10^6 ly according to postulate (1), depending on the actual mass density (D) in the gravitational field. (See section 27 for the expected range of D values.) That diameter has been calculated assuming a spherical rather than a "disklike" distribution of mass in a galaxy and, as such, represents an approximation only for most galaxies. Nevertheless, that approximation is sufficient to conclude that there would be ample space in a typical cosmic void to host the

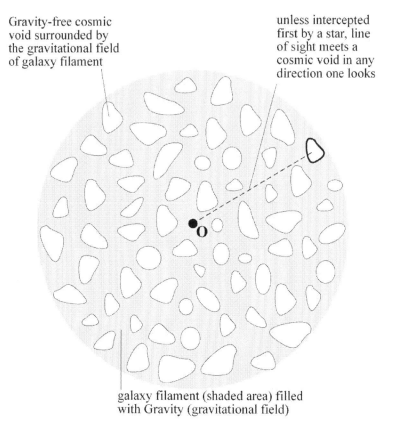

Fig. 25 Structure of the universe comprising gravitational field filled with galaxies (galaxy filament) and Gravity-free cosmic voids containing none or very few galaxies.

gravitational fields of numerous galaxies. Those fields would not be connected to the gravitational field of the galaxy filament. The galaxies inside cosmic voids are called "void galaxies." Gravitational fields of void galaxies could be interconnected—that is, clusters of galaxies can form inside large cosmic voids. A cosmic void can be best visualized as a large, roundish region of the universe delineated by a boundary of the gravitational field of the galaxy filament. It encompasses a large amount of Gravity-free space that can contain individual gravitational fields.

At first glance, one might expect Gravity – no-Gravity boundaries that delineate cosmic voids to be an observational feature of the universe. On a closer examination, however, one has to conclude that there seems to be nothing conspicuous about the boundary of a cosmic void. Consider a star or any other mass object situated in the gravitational field of a supercluster, moving toward a cosmic void's boundary and then entering the void. As the object nears the boundary, the force of gravity acting upon it decreases to a negligible value, and it becomes zero on the opposite side of the boundary. The object feels no significant change in physical interaction and nothing spectacular happens. The boundary does not comprise matter or energy, so it does not generate any observable signals when an object passes through it. The change in the shape of the boundary of the gravitational field would also be impossible to observe.

In section 2, it was assumed that the perfect cosmological principle holds, which requires the universe to be homogeneous and isotropic on a large scale. In section 25, the concept of the force of antigravity was put forward. When accounting for that concept, it appears rational to suggest that the antigravity forces, generated as a result of the execution of the SLT (section 25), are responsible for maintaining the homogeneity and isotropy of the universe on a large scale—that is, the homogeneity and isotropy of the distribution of matter in the universe. This is contrary to the true forces of gravity that act in the opposite direction of increasing that nonhomogeneity by clustering matter together. On a very large scale of an RVU, antigravity forces win. On the scales of a supercluster and smaller, the winners are the true forces of gravity. That no-win eternal battle between the SLT and the true forces of gravity appears to be the most fundamental feature of the dynamics of the macroscopic universe. The no-win battle also means that the entropy in the universe remains constant on a large scale.

39 Cosmic Void and Star Horizons

As pointed out in the preceding section, the distribution of cosmic voids in the universe has to be homogeneous on a large spatial scale so that the cosmological principle is satisfied. In the universe that is spatially infinite, cosmic voids must form a horizon for any observer regardless of the observer's location. The observer is marked by "O" in figure 25. Put otherwise, the line of sight of any observer must eventually intersect a cosmic void in any direction the observer looks in, unless it first intersects a star or a galaxy. (One would not be able to actually see a cosmic void since, as pointed out previously, its boundary comprises neither matter nor energy, and it does not emit or reflect anything.) I call that horizon a "cosmic-void horizon".

The existence of a cosmic-void horizon means that the force of gravity acting on any g-object in the universe is finite. To visualize that, consider a line of sight of the observer located on the planet marked by "O" in figure 25. The planet gravitationally interacts, either directly or indirectly, with any g-object situated along the observer's line of sight up to the location of the cosmic void that intersects that line. An example of such a void is marked with a thicker boundary line in figure 25. The g-objects with centers of gravity located along the observer's line of sight beyond that cosmic void will not exert any forces on the planet.[ff] That is for their gravitational fields along the line of sight are separated from the field in which the planet is embedded. The gravity ranges of those g-objects end on the boundaries of the voids. It follows that owing to the existence of a cosmic-void horizon, the influence of very distant masses on a local gravitational system is nil, if very distant masses are defined as those located behind the cosmic-void horizon of a local gravitational system. It also follows that the forces of gravity acting upon any mass object in the universe are finite in all directions.[gg]

The conclusion that the gravity force acting on the mass object is finite in all directions explains Einstein's proof discussed in section 2. Einstein indeed proved that the universe cannot be spatially infinite if Newton's premise of the infinite range of gravity is to hold. That proof, however, does not imply that the universe cannot be spatially infinite. It merely implies that it cannot be spatially infinite if the range of gravity is infinite. It needs to be noted that a finite range of gravity is not sufficient to resolve Einstein's infinite-force quandary. As discussed in the earlier

---

[ff] The scenario of the g-objects located beyond a cosmic void corresponds well to the Sciama's idea of "very distant masses" discussed in section 24. According to Sciama, nearly all gravitational influence on a local system comes from very distant masses. According to the GPU cosmology, the influence of the very distant masses on a local gravitational system is nil.

[gg] If the range of gravity were infinite—that is, if the universe were completely filled with gravitational field—no Gravity-free regions would exist, and the magnitudes of the forces of gravity acting on a local gravitational system would be infinite in all directions. That is because the infinite number of the centers of gravity of g-objects (i.e., infinite amount of outside masses) would result in the infinite strength of gravitational interactions of the g-object with outside masses.

paragraph, the existence of a cosmic-void horizon that leads to a finite strength of the force of gravity is necessary as well.

Since a cosmic void is formed by the lack of gravitational field, its typical size in the low-galaxy density regions of the universe would be expected to be larger than it would be in the high-density regions. This expectation is in total agreement with the actual findings of the redshift surveys (see, e.g., Lindner et al.[174]).

The other horizon that any observer in the universe should "see" is a star horizon—that is, if one could see radiation emitted at all frequencies, including those emitted by invisible stars/galaxies. The majority of the invisible galaxies are expected to be formed by stars older than WHITE DWARFS. While we cannot actually see radiation outside of the visible-light frequencies, we can detect it using radio telescopes. Infinite time requires that "cold," invisible stars/galaxies are abundant (see section 3). Consequently, the great majority of a star horizon should comprise such stars/galaxies if the universe is infinite in time. That part of the star horizon would be expected to form a dark background of the sky that we observe at night. Parts of the dark sky might also be formed by invisible galaxies "hotter" than the visible-light galaxies (e.g., the so-called "blue" galaxies, in which new stars are still being formed). Note that the existence of invisible galaxies resolves Olbers's paradox, a fact that I alluded to in sections 2 and 10: the vast majority of our star horizon comprises invisible stars/galaxies, the radiation emitted from which is invisible to the human eye so that the night sky is seen to be black.

40 Mass Density in the Universe

There doesn't seem to be any hard, unbiased data that could be used to estimate the mass density in the universe or in the galaxy filament. The expected abundance of invisible stars/galaxies implies that the actual mass density in the universe is much greater than the density that can be estimated by accounting for visible matter only. A total mass of matter in the universe that is larger than the mass of visible matter has often been suggested from the observations of gravitational interactions as they affect the GALAXY ROTATION CURVES (see, e.g., Jeffrey Bennet et al.[175]). Those observations suggest that there is approximately ten times more matter in galaxies than the visible matter. It is normally suggested that the invisible matter comprises dark-matter galactic halos (see, e.g., John

Bahcall et al.,[176] Neta Bahcall and Andrea Kulier[177]). Even that estimate is not unbiased, as it relies on the assumption that Newton's law of gravitation applies in the same form regardless of the distances between interacting objects. As discussed previously, if the range of gravity is finite, that assumption would not be correct, and the actual mass in the galaxy could be significantly higher than ten times the visible mass.

The currently estimated value of the average mass density in the universe is about 1.0×10^{-26} kg m^{-3}. That, of course, is a biased estimate since it represents the CRITICAL MASS DENSITY, which is a concept derived from the standard cosmological model. The critical density has been estimated from the interpretations of sky surveys (e.g., WMAP and BOSS) that suggest that the universe is "flat." That also means that the above estimate relies on all the assumptions that underlie the standard cosmological model: space-time is a physical reality, the big bang did happen, the universe is expanding, the expansion of space is accelerating, etc.

The average mass estimate of 1.0×10^{-26} kg m^{-3} includes dark energy and dark matter, the total amount of which is thought to be about twenty-five times larger than the amount of visible matter. There doesn't seem to be any firm data that would indicate that estimate to be low. However, any estimate of the average mass density in the universe that is based on the observations of visible matter and gravitational interactions could be low. That is because in such estimates, no allowance is made for "dark matter" that can conceivably exist deep in the intergalactic space. (We are unable to observe the effects of gravitational interactions deep in intergalactic space, which, according to the standard model, is assumed to be empty of matter—that is, we are unable to infer the possible existence of matter in most of the universe.[hh]) In other words, any currently available estimate of the average mass density in the universe is biased toward the assumption that deep intergalactic space is empty of matter. I want to emphasize that the infinite-time hypothesis does imply the existence of invisible matter deep in intergalactic space, the amount of which would be expected to exceed the amount of visible matter by a large margin, as pointed out in section 3.

[hh] It is of interest that the existence of a substantial amount of matter deep in intergalactic space, which is currently assumed to be empty of matter, would make the universe more homogeneous in terms of mass distribution.

The foregoing discussion indicates that no currently available estimate of the mass density in the universe is free from some underlying and unconfirmed assumptions. Yet all of those estimates consistently point toward the existence of invisible mass, the amount of which is significantly larger than the amount of visible mass. The invisible mass is called "dark matter" and is normally thought to be non-BARYONIC. (The existence of nonbaryonic matter is an ad hoc hypothesis—no one has ever observed or detected nonbaryonic matter, and no one knows what it could consist of in physical terms.) In the GPU cosmology, dark matter comprises invisible stars/galaxies, which are entirely baryonic. The issue of dark matter is discussed further in section 45 in conjunction with the discussion of the CMB.

41 The Life of a Star

It follows from the infinite-time hypothesis that, on average, for every star that is born, another star must die (section 3). This presents a key characteristic of the GPU cosmological model. In this section, that characteristic will be shown to be physically viable if both the FLH and the insufficient-mass conclusion hold.

In the standard cosmological model, stars are born out of cosmic gas and dust generated in supernova explosions. However, the scenario in which supernovae generate enough gas and dust for the formation of all stars that must be born is not realistic in the infinite-in-time universe. Each supernova would have to lead to the formation of a sufficiently massive star that would result later in another supernova. Otherwise, given the infinite time, there would be no more supernovae. The number of the supernovae and the amount of gas and dust generated in those explosions would decrease with time, as some of it would be used up in the formation of stars having masses smaller than those of supernova-size stars. Smaller mass stars are expected to be abundant. In the infinite-in-time universe, cosmic gas (the basic ingredient of star creation) is continuously created at a rate that is steady on a large time scale. The only known fuel of a quantity sufficient for the creation of the required quantity of cosmic gas appears to be the thermal radiation of stars.

The FLH was used to conclude that the speed of a photon must be constant and independent of the speed of the emitting source only when subject to gravitation. That conclusion was based on the proposition that the photon is a fundamental particle in the sense of the FLH, which means that it cannot have a speed-control mechanism built into it. Recall that the insufficient-mass conclusion implies that Gravity-free regions in the universe have to exist. In those regions, which have been identified with cosmic voids, the speed of a photon (v_{ph}) can be less than c as a result of the collisions with other photons. In the ensuing, I will discuss how this implication relates to the birth of a star.

Consider a region of the universe situated within the galaxy filament, denoted GF, that is completely filled with Gravity. After a collision of two photons in the GF, their speeds, both of which are equal before the collision to c, remain the same because they are subject to gravitation. Sometime after the collision of the photons, they exit region GF because their speeds are maintained at c, which means that no energy of photons accumulates in the GF—that is, in the galaxy filament.

In a cosmic void, the speed of a photon after one or more collisions with other photons can decrease. That means that the total kinetic energy of photons entering a cosmic void can decrease, while their total intrinsic energy would increase.[ii] As the photon speeds are sufficiently reduced in such collisions, the energy of photons can accumulate in cosmic voids in the form of relatively slow-moving, highly energetic photons. It becomes available for the creation of matter (e.g., a collision of two high-energy photons can produce an electron-positron pair). In the process of matter creation, cosmic gas is generated in a cosmic void. It is the basic fuel for the formation of stars and galaxies, the gravitational fields of which may eventually connect to the galaxy filament. As the density of cosmic gas in a star or a void galaxy that is being formed increases, the frequency of photon collisions increases, which results in an increase in the rate of the accumulation of photon energy. Dense clouds of cosmic gas can form a continuous gravitational field that may connect to the galaxy filament. Note that the cold-gas accretion in void galaxies has been discovered, as a "most tantalizing finding," from actual observations of void galaxies

---

[ii] In section 29 I pointed out that the total energy of a photon can be considered to be either kinetic energy or intrinsic energy. That is no longer valid in a Gravity-free space, as the speed of a photon can vary. For instance, if the speed of photon is reduced to zero (in the frame of reference of an observer), its kinetic energy becomes zero, while its intrinsic energy is nonzero and proportional to its frequency, in accordance with equation (22).

(R. van de Weygaert et al.[178]). This scenario of the creation of cosmic gas out of electromagnetic radiation explains well the relatively high rate of the accretion of cosmic gas in cosmic voids.

Actual astronomic observations appear to support the scenario of star/galaxy formation in cosmic voids. First, the observations show that the majority of galaxies inside cosmic voids are relatively young—that is, their spectra are "bluer" than those of the typical filament galaxies, as discovered by Rojas et al.[179] and others. That is consistent with the scenario of galaxies being born in cosmic voids. Second, when formed in the Gravity-free environment of a cosmic void by the creation of mass at an extremely low rates, galaxies would be expected to acquire regular shapes such as spiral or elliptical—that is, disklike configurations—due to the "self-contained" gravitational effect. The idea that galaxies formed in cosmic voids should have disklike shapes is in good agreement with astronomic observations, which show no irregular void galaxies present in cosmic voids (Goldberg et al.[180]). Third, the majority of void galaxies are rich in cosmic gas (Kreckel et al.[181]), which is consistent with the scenario of cosmic-gas creation out of radiation in cosmic voids. The cosmic gas is the basic fuel for star/galaxy creation.

Astronomic observations also show that, in cosmic voids, the rate of galaxy formation is higher and that the fraction of PASSIVE GALAXIES is lower than in the galaxy filament (e.g., E. Ricciardelli et al.[182]). Both those observations can be predicted from the GPU cosmological model in which cosmic gas (i.e., the basic fuel for the creation of stars) is generated primarily in cosmic voids.

As previously noted, in the universe that is infinite in time, each star will die by emitting its entire energy, except for those stars that die by explosive emissions of energy, such as supernovae. A nonexplosive emission of the entire energy of a star must take an extremely long time. That is to be expected, as the birth of a star out of radiation must take an extremely long time as well. The energy of a typical photon in a cosmic void has to increase, due to collisions with other photons, to at least the gamma level, and then it has to encounter another high-energy photon and collide with it so that an electron-positron pair and other particles can be created. This points to an extremely long matter creation process.

According to the standard cosmological model, the oldest stars in the universe are between about twelve and thirteen billion years old. In the infinite-in-time universe, much older stars have to exist. The nuclear fusion would have ceased is those stars and they would be expected to emit radiation at very low temperatures, with no visible light emissions.

The age of the sun is estimated to be about 4.6 billion years. If that is about correct, and if the sun was actually formed out of radiation, then the lifespan of a star would be expected to be much longer than the age of the universe implied by the standard model (about 13.8 billion years). That may seem to present a far-fetched suggestion. Yet, an age of more than 13.8 billion years is still unimaginably short when compared to infinity. When researching cosmology, one should not humanize nature by thinking of the length of a star's life from the human's perspective. Humans' perception of time stems from the length of a day and, what follows, from the length of a year. Yet the galactic year, for instance, is about eleven orders of magnitude longer. There is nothing really long about, say, 1,000,000,000 billion years in comparison with 13.8 billion years. From the perspective of infinite time, those two time periods are close.

42 Perpetuum Mobile

If one thinks of the infinite-in-time universe as a machine, one will soon realize that it is the only perpetual-motion machine that can exist. I suspect that most physicists would find the existence of a perpetual-motion machine objectionable. Yet the mere inference that the universe is infinite in time is sufficient to conclude that the universe is indeed such a machine. As pointed out in section 3, the entropy in the infinite-in-time universe has to be constant and not increasing. The unambiguous fact that we do not observe infinite entropy in the universe means that, on a large scale, the entropy in the universe has not been increasing through the past infinity of time.[jj]

In the infinite-in-time universe—which has to be spatially infinite as well—the energy of thermal radiation emitted from a star is either

---

[jj] In a representative volume of the universe that is infinite in time, entropy cannot increase or decrease. It must remain constant. There is a comfort in an infinite-in-time-universe universe: the heat death of the universe will not happen. This comfort, however, is somewhat offset by the following discomfort: in the universe that is infinite in time, any product of evolution, including any civilization, must eventually perish to satisfy the perfect cosmological principle.

absorbed by other stars that form the emitting star's horizon or used in the formation of new stars. A star that is formed out of radiation decreases the entropy (the disorder) in the universe. A star that absorbs radiation decreases the entropy as well. Why is it, then, that according to the standard cosmological model, the entropy in the universe increases? The answer to this question appears to be: some of the energy of thermal radiation emitted by stars is never absorbed by other stars or used in the formation of new stars. Consequently, there is more energy dispersion than energy accumulation in the universe, which leads to an increase in the disorder in the universe—that is, to an increase in the entropy of the universe. In the infinite-in-time universe, a balance of thermal radiation in an RVU makes the entropy constant with time, owing to the perpetual emission and absorption of radiation:

amount of radiation emitted by stars
=
amount of radiation absorbed by stars
+
amount of radiation used in the creation of mass (stars)

The first term of this equation represents an increase in the disorder (entropy) in the universe, while the second and third terms represent the corresponding decreases in the disorder. Therefore, the above equation shows that the birth and the death of stars result in a perfectly reversible process on a large time and space scale.

43 Distribution of Galaxies in Galaxy Filament

Astronomic observations indicate that galaxies are grouped in gravitationally bound clusters. If it actually is gravitation that forms the clusters of galaxies, then it is reasonable to expect that typical clusters should have a roundish shape just like most galaxies do—that is, the structure of a galaxy cluster would be expected to be similar to the structure of a galaxy. The roundish shapes of many galaxy clusters have been reported from observations (e.g., Pascal de Theije et al.[183]). Also, it can be expected that the number density of galaxies should increase toward the center of a cluster just as the number density of stars increases toward the center of a galaxy. That is how gravitation is expected to work over long time: it pulls all the stars that form a galaxy together and pulls all the galaxies that form a cluster of galaxies together.

Now, if there are many more invisible galaxies than visible galaxies, and if the visible galaxies are spread out among invisible galaxies in an approximately uniform pattern in accord with the cosmological principle, then it is reasonable to expect that the concentrations of visible galaxies would be higher near the centers of galaxy clusters. In other words, the high-density regions of visible galaxies are expected to be located near the gravity centers of galaxy clusters—that is, if there are more galaxies in a region of the galaxy filament, then there should be, on average, proportionally more visible galaxies in that region. This could provide a rational explanation of the observed clustering of visible galaxies.[kk]

44 Cosmological Redshifts

The expansion of the universe assumed in the standard cosmological model is based on the galaxy redshifts interpreted as the Doppler shifts.[ll] On the other hand, the expansion of space is not possible according to the infinite-time hypothesis and the resulting spatial infinity of the universe (section 3). Regardless, there appears to be no question that starlight beams emitted from galaxies should be subject to the Doppler shift—that is, their spectra should be redshifted or blueshifted, depending on the direction of the radial motion of a galaxy in an observer's frame of reference. Without that property of radiation beams, GPS wouldn't work, and police wouldn't be able to issue speeding tickets. If radiation beams comprised particles (photons) rather than waves, they would also be subject to redshifts or blueshifts, the amount of which would depend on

---

[kk] The clustering of galaxies is often described using the concept of hierarchical clustering, which means that larger structures in the universe are formed by the merging of smaller structures. That description can also apply to a universe in which invisible galaxies are abundant, as in the GPU cosmological model.

[ll] In the standard model, the shift due to the expansion of space rather than the classical Doppler shift is actually assumed. In the former case, the wavelength of emitted photon is determined by the speed of the expansion of space between a galaxy and the observer (galaxies are not moving in relation to space). In the latter case, the wavelength is determined by the speed of the emitting galaxy receding from the observer—that is, galaxies are moving. Those are equivalent concepts, and the more familiar Doppler-shift concept is used in the discussions.

the speed of the beam's emitting source. That is for the Doppler-shift effect applies as much to waves as to other periodic events. For instance, it applies to a stream of particles that periodically arrive at an observer's location. However, supposing that starlight beams comprise particles (photons) rather than waves leads to a more satisfactory interpretation of redshifts, which also supports the premise of a nonexpanding universe. In the following discussion, the photons that form a starlight beam will be submitted not to lose any energies when traveling through the universe. This is contrary to the classic tired-light interpretations of redshifts, in which the redshifts are caused by energy losses in starlight beams.

The assumed unique relationship between a cosmological redshift of starlight beam and the radial speed of the beam's emitting galaxy, which I will call "the Doppler assumption," is the foundation of the standard cosmological model. In this regard, it is critical to realize that the Doppler assumption incorporates another assumption that appears to be rather unreasonable: the characteristics of a starlight beam traveling over distances as vast as billions of light years remain completely unaffected in spite of the expected gravitational interactions of the photons that form the beam with other energy and/or matter. The assumption incorporated in the standard cosmological model is that redshifts are entirely due to the receding speeds of starlight-emitting galaxies.

Many scientists have suggested that the Doppler effect could be only partially responsible for the observed redshifts. That suggestion has often been made with reference to the redshifts of QUASARS (see, e.g., Jayant Narlikar,[184] Halton Arp et al.,[185] Arp[186]). It is also worth noting that Jean-Claude Pecker did identify some problems with the Doppler's interpretation of cosmological redshifts with specific reference to the observed spectra of the sun, binary stars and close-by galaxies.[187]

To appreciate how a starlight beam can be affected while traveling over the vast distances of the universe, one needs to realize first that the classic Doppler assumption, as used in the standard-model interpretation of cosmological redshifts, is based on Maxwell's concept of a light beam that comprises a smooth continuous wave. That wave is not subject to gravitational interactions. Today, it is known from quantum physics that a light beam can exhibit a wavelike behavior even if the photons have been emitted one at a time, which, clearly, is not a characteristic of a wave. As noted in footnote f (section 1), any periodic phenomenon can be described using wave mathematics, which does not necessarily mean that the phenomenon is a wave. It seems that the word "wavelike" should be taken literally when examining the behavior of particles.

Consistent with Feynman's insight,[188] I assume that a starlight beam comprises particles (photons). Also, consistent with the implication of equation (3), photons in gravitational field are subject to gravitational interactions. It can be concluded, therefore, that the motion of each of the photons in a starlight beam can be affected differently by gravitational interactions with cosmic objects when traveling through the gravitational field of the galaxy filament.

Suppose that a starlight beam arrives at the earth after traveling for five billion years following its emission. The photons that form the beam have traveled over a distance of about 5×10^{22} km before arriving at the earth. According to the Doppler assumption, while the photons travelled over that enormous distance, they have not interacted in any way with anything that might affect the beam's redshift. This also means that each photon has traveled along an exactly rectilinear pathway, as a change in the direction of a photon's pathway would be expected to affect the amount of redshift. (I will discuss this inference in more detail shortly.) Such a scenario does not seem to be very credible since the photons in a starlight beam would be subject to gravitational interactions whenever travelling through a gravitational field.

As discussed above, the photons in a starlight beam are expected to gravitationally interact with other cosmic objects such as galaxies, stars, and cosmic gas and dust. Those interactions are expected to result in (slight) changes to the directions of some or all of the photon pathways without any losses in their energies/momenta. Nonetheless, losses in the photon momenta in the beam direction are expected. As the redshift can be considered to be a measure of the photon momenta in the direction of the beam, the generation of redshift due to gravitational interactions of the photons is expected.

A starlight beam runs toward the earth. It comprises photons. As discussed above, the directions of some or all of the photons' pathways are expected to change as the result of the gravitational interactions of photons. Consequently, the component velocities of the photons in the beam direction are expected to be *reduced*. Those reductions would be expected to increase in proportion to the distance traveled—that is, in proportion to the amount of gravitational interactions that the photons have been subjected to. It follows that as the traveling distance of a starlight beam increases, the component velocities in the direction of the beam of some or all of the photons would be expected to decrease. Hence, there would be associated reductions in the components of the photon momenta in the beam direction. The "wavelengths" of the photons would be expected to increase, which would correspond to an increase in the measured redshift of the beam. That effect is analogous to observing the starlight emitted from a galaxy that is moving at an angle from the line connecting the galaxy and the observer. The larger the angle, the larger redshift would be observed, with the energy of the beam conserved.

To appreciate how the above scenario may relate to cosmological redshifts, consider the redshift effect using a highway analogy. There is a one-way multilane highway with moving cars evenly spaced, and each lane is moving at a different but constant speed. An observer stationary in relation to the highway is counting the passing cars during equal time intervals and sees the cars coming in waves: sometimes cars from all of the lanes arrive during the time interval, sometimes fewer cars arrive during the interval, and sometimes none arrive. The traffic's wavelength is constant. If the car speeds in some lanes are reduced, the wavelength becomes longer. (This corresponds to a reduction in the components of some photons' momenta in the direction of a starlight beam.) If those speeds are increased, the wavelength becomes shorter. (The components of some photons' momenta in the direction of a beam increase.) If the observer starts moving in the direction of the highway traffic, the wavelength becomes longer, and vice versa. The wavelike behavior of starlight beams is a consequence of the corpuscular nature of light according to the highway analogy: if all the cars were traveling literally "bumper to bumper," the traffic's momentum would be arriving smoothly rather than in waves. There is nothing new shown in the above highway analogy. It is just a re-statement of the applicability of the Doppler-shift effect to periodic events. The advantage of presenting this analogy here is that it allows for explaining how the variable property of the photons—that is, the momenta of photons in the beam direction—can affect the cosmological redshift of a starlight beam, *without* denying the applicability of the Doppler-shift effect.

In summary, with an increase in the travel distance of a starlight beam, the components of some or all of the photon velocities in the beam direction (and the corresponding momentum components) are expected to be reduced due to gravitational interactions of photons with other cosmic objects. Accordingly, the wavelength of the beam is expected to increase—that is, the beam is expected to be redshifted. Such reductions in the components of photons' momenta should be reflected in the electromagnetic spectra of light beams. A change in the components of the photons' momenta can only result in a redshift effect, but not in a blueshift effect. That is because, on the average, the components of the photons' momenta in the beam direction would not be expected to increase with the distance traveled. The reduction of the components of photon momenta will be referred to as the RCoPM effect.

In the proposed explanation of cosmological redshifts, a relation between the speed of the emitting source and the amount of the redshift (i.e., the Doppler-shift effect) is allowed for, just like in the standard cosmological model. This has been illustrated by the highway analogy, in which the wavelength of traffic depends on the speed of an observer moving in relation to the traffic. However, a change in the redshift would also be expected even if the source of a starlight beam is *stationary* in the observer's frame of reference. The directions of photon pathways would still be changed by gravitational interactions after the emission. Owing to the nearly homogeneous distribution of matter in the galaxy filament, that change should be approximately *linear* with the distance traveled as long as the starlight beam runs through a continuous gravitational field. (Simply, twice the distance travelled, approximately twice the amount of the RCoPM effect would be expected).

However, the gravitational field extending along a starlight beam pathway that covers large cosmic distances is occasionally intercepted by cosmic voids—that is, by Gravity-free regions of the universe (section 38). As discussed in section 39, each observer is surrounded by his/her cosmic void horizon, beyond which a starlight beam has to travel through some cosmic voids. In the voids, the beam would not be subject to any RCoPM effects. It follows that for large cosmic distances, the distance vs. redshift relationship can be expected to be nonlinear, with the redshift increasing at a slower rate than the distance. This is consistent with the actual redshift observations reported by S. Perlmutter et al.[189] and others.

According to the proposed interpretation of redshifts, the effect of the RCoPM is expected to overwhelm the Doppler shift for starlight beams emitted from far away galaxies. While those galaxies could be moving toward us or away from us with some "peculiar" velocities, we would always observe galaxy redshifts. For close-by galaxies, the RCoPM effect could be too weak to overwhelm the Doppler-shift effect, so that we might observe those galaxies being redshifted or blueshifted. This is consistent with Hubble's original findings. He found that, out of the twenty four galaxies that he studied, the spectra of the four closest-to-the earth galaxies were blueshifted (except for the Magellanic Clouds), which indicated that those were moving toward us. Hubble also found a fifth blueshifted galaxy located farther away, which had a very small blueshift. Those blueshift observations are consistent with the prediction that the RCoPM effect overwhelms the Doppler shift primarily at larger distances, at which the amounts of gravitational interactions that photons have been subjected to are significant. If the proposed explanation of redshifts is correct, one can expect that the larger the distance to a galaxy is, the smaller is the chance to see its starlight being blueshifted, which is consistent with Hubble's findings.

The proposed explanation of cosmological redshifts does not mean that those redshifts have to be due to the RCoPM effect and the speed of the emitting galaxy only. There is a strong evidence that at least some galaxies exhibit intrinsic redshifts (Arp,[190] Lopez-Corredoira and Carlos M. Gutiérrez[191]). The intrinsic redshifts can significantly contribute to the redshifts associated with the RCoPM effect and the speed of the emitting galaxy. Recall that the effect of the RCoPM on the observed frequency of radiation beams is expected to be negligible at very short distances—for instance, at the GPS distances. The above evidence questions the uniqueness of the redshift vs. distance relationship. That relationship does not need to be the same for all starlight beams. It is expected to depend on the mass of the emitting star/galaxy, which determines the amount of gravitational redshift of the emitted beam. In most cases, however, that dependence would be expected to be negligible.

Let me elaborate on the gravitational redshift effect. The majority of stars/galaxies observed using optical telescopes emit radiation at similar emitting mass levels. Thus, a similar redshift vs. distance relationship is expected to hold for those stars/galaxies, and it does. However, that relationship needs not be accurate for all cosmic objects. As an example, for quasars, which are extremely massive objects, that relationship could be significantly different. Photons emitted from a quasar would be subject to a very high gravitational redshift and thus lose a relatively high amount of energy shortly after the emission. (Light would get 'tired' fast shortly after the emission.) That means that some quasars could be located closer to the earth than has been estimated using the standard interpretation of cosmological redshifts. This has been long advocated by Arp, Fulton, Carosati[192] and others. A redshift caused by the gravitational interaction of photons with the emitting objects presents just an example of possible intrinsic redshifts. There are other potential causes of such redshifts that could be associated with the properties of emitting objects.

Let me now neglect the intrinsic redshift effect and consider two starlight beams emitted from galaxies located at the same distance from an observer. The beams were emitted at different frequencies. According to Hubble's law of redshifts, the observed redshifts should be the same in both beams (Hubble's redshift vs. distance relation does not depend on photon frequency). The proposed explanation of measured redshifts leads to the same conclusion. The photons in the two beams have been subject to about the same gravitational influence, which suggests similar RCoPM effects, that is, similar deflections of photon pathways. The deflections of pathways caused by photon interactions with massive objects are not expected to appreciably depend on photon frequencies (section 19).

A more suitable explanation of cosmological redshifts is, of course, possible. It seems to me that the strength of the proposed explanation is that it avoids the assumption that a light beam traveling over billions of light-years is not affected in any way by other energy and/or matter. In this regard, I want to emphasize that it has been experimentally proven that starlight beams are indeed affected by gravitational interactions with massive cosmic objects (e.g., with stars, galaxies, or clusters of galaxies). Those proofs involved the observations of the effects of gravitational lensing and the deflection of a light beam passing next to the sun.

It seems that there are three primary receding-galaxy arguments that could be stated against the infinite-in-time universe. The assumption of Doppler shift being the only cause of cosmological redshifts could be the leading argument against such a universe, since it directly suggests that galaxies are receding and, consequently, the age of the universe has to be finite. The second primary argument could be purely theological: the universe cannot be infinite in time since in such a universe, there is no room for an act of creation. The third primary argument could be that no stars older than about fourteen billion years have been observed. The lack of evidence that would support the first argument was discussed in section 1, and an alternative interpretation of redshifts was put forward in this section. The second argument is not scientific. It relies on faith and as such cannot be debated. The third argument, which relates to dark matter and the origin of CMB, is addressed in the following section.

45 Cosmic Microwave Background

Next to the Doppler-shift interpretation of cosmological redshifts, CMB is widely considered to be the most important evidence supporting the standard cosmological model. Nevertheless, the merits of that belief are frequently questioned (e.g., Assis and M. C. D. Neves[193]). Geoffrey Burbidge[194] specifically points out that the standard cosmological model cannot explain the CMB. Similar to the problem with the Doppler-shift interpretation of cosmological redshifts, CMB being the relic radiation from the big bang presents an assumption only, which has no supporting direct evidence. This, of course, does not mean that the big bang, the early inflation of the universe, the photon decoupling, etc., could not happen. It means, however, that another genesis of CMB is possible.

In the GPU cosmology, the amount of (baryonic) dark matter is expected to be significantly larger than the amount of matter estimated from astronomic observations, including the observations of gravitational interactions. As discussed in section 3, the majority of stars/galaxies in the universe are expected to be very old and very cold (VO&VC). Those and other invisible to human eye galaxies are expected to be abundant deep in intergalactic space, and to form the majority of our star horizon. It should be noted that no observations of the gravitational effects of dark matter deep in intergalactic space are currently possible, except for some observations of the effects of gravitational lensing. The invisible galaxies and the CMB appear to present physically coherent candidates for dark matter and the radiation of dark matter, respectively. While I am unable to prove that the CMB is the radiation emitted from VO&VC galaxies, I want to emphasize that, in principle, it is possible to obtain such proof.[mm] It could be obtained by sending a cosmic probe designed to detect and map mass objects deep in the intergalactic space. Of course, this will not happen any time soon. Regardless, improvements in the resolution of radio telescopes used to detect CMB could, in the foreseeable future, let us identify (the better phrase is "let us see") the VO&VC galaxies.

In discussions claiming that CMB is the relic radiation from the big bang, it is rarely mentioned that CMB represents only the most dominant part of cosmic background radiation. As indicated in section 3, invisible galaxies that emit radiation at frequencies other than the CMB range of frequencies are expected to exist, and they appear to exist. In this regard, the existence of cosmic infrared background and X-ray background are well documented, as is the existence of radiation emitted at many other "invisible" photon frequencies. One needs to keep in mind, however, that some of the observed radiation could be of nonthermal origin.

At first sight the VO&VC galaxies correspond well to the MACHOs (massive astrophysical compact halo objects) often considered to be a top candidate for dark matter. However, there is a fundamental difference between the two dark matter proposals. The VO&VC stars/galaxies are expected to emit radiation with energies (frequencies) determined by the surface temperatures of the emitting stars. This is consistent with the well-established understanding of the behavior of matter and energy. MACHOs, on the other hand, are expected to emit no radiation at all or to emit very faint, undetectable radiation only. Therefore, this presents a

---

[mm] In the standard cosmological model, it is normally assumed that dark matter does not emit radiation. That assumption is based on the assertion that we do not detect dark matter's radiation. As far as I know, this is the only background to that assumption. (A more rational background would be: dark matter does not emit radiation because of its special properties, such as ... .) The assumption that the CMB is not radiation emitted by dark matter is a necessary characteristic of the standard model, according to which the CMB radiation was generated at the time of the last scattering. There appears to be no other reason than the assumption that the big bang happened that would eliminate the origin of CMB from being the dark matter's radiation. Put otherwise, it has to be assumed in the standard model that the big bang did happen (which cannot ever be confirmed) and that radiation-free matter exists (the possibility of the existence of such matter has never been confirmed). In the GPU cosmology, the existence of VO&VC stars/galaxies (dark matter) can, in principle, be confirmed.

proposal that hypothesizes the existence of matter that has never been observed, while the possibility of the existence of such matter has never been explained. For instance, it has never been explained if dark matter would be made of ordinary hadrons and leptons that, in turn, would form atoms and, if so, why those atoms would not emit radiation. That lack of explanation, it seems, makes the proposal that the dark matter comprises VO&VC stars/galaxies more rational since no existence of an unknown kind of matter has to be postulated. As a result, the proposal that CMB is radiation emitted by dark matter appears to be rational as well.

If the CMB is indeed the thermal radiation emitted by VO&VC galaxies, then small variations in the observed frequencies of the CMB photons would be expected. This is because the CMB photons would be arriving at earth laboratories in the form of radiation beams rather than in the form of a smoothed-out radiation as in the case of the relic radiation from the big bang. The sources of CMB—that is, the VO&VC stars—would be located at various distances from the earth. As a result, the CMB radiation beams would be traveling over various distances, and they would be subject to cosmological redshifts. For a given distance, the redshifts observed in a CMB beam are expected to be approximately the same as the redshifts in the visible-light beams. (As discussed in section 44, the deflections of the pathways of photons emitted at different frequencies are expected to be similar.) Since the frequencies of CMB radiation beams at the time of emission would be very low, the observed differences in the frequencies of the CMB radiation beams emitted at various distances would be very small.

The perfect cosmological principle requires that the temperature in intergalactic space (T_{sp}), normally referred to as the cosmic background temperature, is essentially constant and uniform throughout the universe. That temperature is set by the incoming radiation, with the CMB beams being the dominant radiation. The surface temperatures of coldest stars are expected to be about the same as the cosmic background temperature, which means that the coldest stars would be in thermal equilibrium with their surroundings. Note that there should be no stars with surface temperatures lower than T_{sp}. This is because a decrease in the surface temperature of a star to below T_{sp} would mean a decrease in the thermal disorder in the universe.

46 Summary of the GPU Cosmology

Let me now summarize what this book has been all about. It has been about the cosmology of an infinite-in-time universe and the physics underlying the concept of such a universe. The physics aspects discussed in this book were those intended to demonstrate the viability of the GPU cosmology. The GPU cosmology has been founded on three hypotheses:

(i) The perfect cosmological principle, which implies the infinite-time hypothesis, holds.
(ii) Fundamental laws of nature (physics) do exist independent of the human mind.
(iii) The (volumetric) range of the gravitational interaction of a g-object is proportional to its intrinsic mass.

Additional assumptions were introduced in order to derive a law of gravitation, which turned out to be Newton's law of gravitation modified to account for a finite range of gravity. A quantitative law of gravitation (an approximate mathematical description of gravitational interactions) is not necessary to formulate a cosmological model at the conceptual level. Nonetheless, the modified version of Newton's law helped support the GPU cosmological model by showing that: the simplicity-of-execution principle is rational and can be used to formulate physical laws; the force of gravity depends on the average mass density in the gravitational field; and the principle of equivalence—while extremely accurate under typical experimental conditions—is approximate only.

Note that it would be entirely appropriate to replace the physical law of gravitation proposed in section 22 with another physical law for the purpose of the GPU cosmological model—for instance, to replace it with general relativity—as long as one recognizes that general relativity is a mathematical model of gravitation and that curved space-time presents a mathematical framework rather than a physical reality. (A modification of the equations of general relativity would be required to account for a finite range of gravity and the temporal infinity of the universe.)

The GPU cosmological model is simple and verifiable in principle. It has been derived from the three hypotheses quoted above. The primary features of that model are listed below. Proving any of those features incorrect could defeat or significantly undermine the GPU model.

- The universe is infinite in time and space.
- The universe is homogeneous and isotropic on a large scale.
- The range of gravitational interactions is finite.
- The gravity forces acting on any object in the universe are finite.
- The entropy of the universe is constant on a large scale.
- Photons are subject to gravitational interactions just as intrinsic mass is. However, a physical law that describes the gravitational interactions between mass objects is not expected to be adequate for the description of the interactions of photons with mass objects.
- The speed of photon (light) must be constant and independent of the speed of the emitting source only when subject to gravitation.
- Gravity-free regions identified with cosmic voids have to exist in the infinite-in-time universe. They are uniformly embedded in the spatially infinite gravitational field that forms the galaxy filament.
- The large-scale homogeneity and isotropy of the distribution of matter in the universe appears to be the result of the execution of the SLT with respect to the nonequilibrium in mass distribution.
- All observers in the universe are surrounded by two horizons: a star horizon and a cosmic-void horizon.
- The energy of thermal radiation can accumulate in cosmic voids, leading to the creation of matter and the formation of stars/galaxies, which may eventually enter the galaxy filament. (Stars and galaxies can also form in the galaxy filament.)
- Most stars die by emitting their entire energies. The lifespan of a star is expected to be many times longer than the age of the universe estimated based on the standard cosmological model.

As demonstrated throughout this chapter, the GPU cosmology is supported not only by theoretical considerations but, also, by satisfactory predictions of some of the key astronomic observations. The most noteworthy of those predictions include:

- The size of a cosmic void in a low density region of the universe is typically larger than in a high density region.
- The amount of mass in the universe is significantly larger than the amount of mass that has been observed, accounting also for the observations of gravitational interactions.

- A typical cosmic void contains a higher percentage of younger (bluer) galaxies, as compared with the galaxy filament.
- A typical cosmic void is characterized by a relatively high rate of the accretion of cosmic gas.
- Cosmic voids exhibit a higher rate of star formation as compared with the galaxy filament.
- Void galaxies have regular (either elliptical or spiral) shapes, which is consistent with their very slow formation in a Gravity-free space.
- At larger distances, the distance – redshift relationship is nonlinear, with the redshift increasing at a slower rate than the distance.
- The cosmological redshift effect, which is suggested to be primarily the result of gravitational interactions of photons with other cosmic objects, overwhelms the Doppler effect at large distances. Hence, it is only the starlight that has been emitted in a close-by galaxy that can be blueshifted.
- Measured redshifts do not substantially depend on the frequency of the emitted photons.
- CMB, as the thermal radiation emitted from very cold and very old galaxies, is expected to exist.

A modern physicist would likely say at this point that "those are not very meaningful predictions as they are not quantitative." My comment in this regard is that the quantitative predictions of any cosmological model have to depend on the mathematical framework that is put over the underlying physical concept. As an example, Newton's very simple equation (9) allows for making highly accurate quantitative predictions of gravitational interactions. Replacing equation (9) with the ten partial differential equations of general relativity, makes such predictions even more accurate. In my mind, this does not mean that Einstein's concept of gravity is more meaningful than, or superior to Newton's concept.

The evolution of a star, from its birth through its death, appears to be the most essential aspect of the GPU cosmology that needs to be addressed by analytical studies and modeling. At this time there is little known about the evolution of stars in which nuclear fusion has ceased—that is, the stars that are older than white dwarfs. Having such a model would allow for an assessment of the amount of dark matter (i.e., the mass density of invisible galaxies) and the typical lifetime of a star.

APPENDIX A: Glossary of Terms and Concepts

Action at a distance is the mode of physical interaction between objects according to which an object can be affected by another object without being physically "touched" by the other object. This idea constitutes a nonlocal mode of physical interaction. The action-at-a-distance mode of interaction is the alternative to the field mode of physical interaction, as explained in section 9.

Ad hoc hypothesis refers to a hypothesis added to a theory in order to save it from being falsified. Most often, an ad hoc hypothesis is a hypothesis added to a theory in order to dismiss an actual finding that implies a failure of the theory.

Baryonic matter is ordinary matter comprising protons and neutrons. (In astrophysics, the "baryonic matter" is meant to also include electrons and neutrinos, which, in fact, are leptons and not baryons.) For the purpose of this book, it is not essential to know what baryons and leptons are. It is essential to realize, however, that no matter other than baryonic has ever been observed. Therefore, as of this time, any claim that non-baryonic matter exists, presents an ad hoc hypothesis.

Binary star refers to a star system in which two stars are orbiting around their center of mass.

Blackbody refers to a body that completely absorbs all incoming radiation regardless of its energy or direction. A blackbody emits electromagnetic radiation with a spectrum that depends on the blackbody's surface temperature. (A "spectrum" is the energy intensity distribution of electromagnetic radiation.)

Bulge of galaxy is the central region of a galaxy in which the majority of the galaxy's stars are located.

Center of mass refers to the average position of an array of masses calculated from the weighted average of distances and masses. A center of mass is the same as the more familiar center of gravity if the same (local) gravitational acceleration exerted at each location of the array's constituents.

Cosmic gas and dust are the matter spread over the interstellar or intergalactic space. Cosmic gas primarily comprises hydrogen and helium, while cosmic dust (solid particles) primarily comprise carbon, silicon, and oxygen. In both the GPU and the standard models, stars are formed out of cosmic gas or gas and dust.

Cosmic microwave background (CMB) is the highly uniform electromagnetic radiation that is detected using radio telescopes. It comes uniformly from each direction in the universe, showing no identifiable source(s). It is a radiation that comprises very low-energy (very "cold") photons. In the standard cosmological model, the CMB is assumed to be the relic radiation from the big-bang event. It is the radiation that was left over after the universe cooled down, following the big bang, sufficient for protons, neutrons, and electrons to form neutral atoms that would no longer absorb radiation. By now, that radiation has very substantially cooled down as a result of the expansion of space, which has led to a decrease in the energy (frequency) of photons.

Cosmic voids are vast discrete regions of the universe that are typically of a roughly spherical shape and contain very few or no galaxies. Cosmic voids occupy more than half the volume of the observable universe and have typical dimensions in the order of 10–150 Mpc (megaparsecs), with an average diameter of about 30 Mpc. The boundaries of cosmic voids are formed by the boundaries of galaxy filament, which is comparatively rich in galaxies.

Cosmological redshift refers to the shift in absorption lines observed in the optical spectrum of starlight and other radiation emitted by stars/galaxies, as measured by an observer on the earth or at any other location in the universe. The pattern of absorption lines at the time that radiation is emitted is known because of the knowledge of the spectra of various chemical compounds from experiments on the earth. In the starlight beams, those lines are shifted toward the spectra of lower-energy photons (lower frequencies of photons). The shift is called a "redshift" because of the examinations of the visible-light spectrum, which indicate that the frequencies of starlight beams are shifted toward the red end of the visible light spectrum. For radiation that cannot be seen by the human eye, "redshift" means a shift of absorption lines toward the spectra of lower-energy photons. Note that *gravitational* redshifts differ from cosmological redshifts. They are generated as a result of gravitational interactions between emitted photons and massive objects such as planets or stars.

Appendix A: Glossary of Terms and Concepts

Critical mass (energy) density in the universe is the mass density incorporated in the standard cosmological model. If the actual mass density in the universe equals the critical mass density, the expansion of the universe will eventually cease. Measurements of the cosmic microwave background, if interpreted according to the standard model, indicate that the actual mass density equals the critical density within a 0.5% margin of error).

Dark energy is a hypothetical energy that uniformly permeates all of space-time. The form of the dark energy is unknown. Its existence is necessary to explain the current accelerated expansion of the universe implied by observations of supernovas and the standard cosmological model. Dark energy has never been detected.

Dark matter is a hypothetical matter that exists in the universe, clumped in galaxies. It cannot be seen or detected. The existence of dark matter is inferred from the observations of the gravitational influence of "something" on the visible matter. While the normal matter primarily comprises protons and neutrons, the dark matter, or, at least, its portion, is believed to be nonbaryonic. It is also believed that nonbaryonic matter does not emit radiation and as such cannot be detected.

The Doppler (shift) effect is the change in the frequency of a wave as measured by an observer moving with respect to the source of the wave. If the source of a wave is moving toward the observer, the times between the arrivals of the wave crests appear to the observer to be shorter than they appear if the source were stationary. That means that the frequency of the wave appears to be higher. The Doppler effect is best documented by observations and analyses of sound waves.

Early inflation of the universe, which is a trait of the standard cosmological model, refers to a time period shortly after the big-bang explosion during which an accelerated expansion of space took place. Between 10^{-36} and 10^{-32} seconds after the big bang, the universe expanded (increased in size) by a factor of 10^{26}. This ad hoc hypothesis was introduced by Alan Guth in 1981 to resolve the flatness and horizon problems that had arisen in conjunction with the original big-bang theory.

Einstein's proof was stated as follows: "According to the theory of Newton, the number of 'lines of force' which come from infinity and terminate in a mass m is proportional to the mass m. If, on the average, the mass density p[0] is constant throughout the universe, then a sphere of volume V will enclose the average mass p[0]V. Thus the number of lines of force passing through the surface F of the sphere into its interior is proportional to p[0] V. For unit area of the surface of the sphere the number of lines of force which enters the sphere is thus proportional to p[0] V/F or to p[0]R. Hence the intensity of the field at the surface would ultimately become infinite with increasing radius R of the sphere, which is impossible."

Electromagnetic radiation is a process in which all objects that have temperatures above absolute zero emit electromagnetic radiation (with the exception of a few fundamental particles). The emitted radiation—that is, the energy comprising photons—originates primarily as thermal radiation. Upon its emission, thermal radiation is converted into electromagnetic radiation. Electromagnetic radiation can also be generated in other electromagnetic processes. For instance, it can be the result of a nuclear explosion, in which intrinsic mass-energy is converted into radiation energy. In this book, the term "electromagnetic radiation" always means thermal radiation. The sun emits radiation in the visible-light frequency range (photons have relatively high energies). An apple or a stone at room temperature emits radiation in the infrared range, with the emitted photons having much lower energies. The range of emitted photon energies depends on the surface temperature of the emitting object. Note that a photon's so-called "frequency" is not really a frequency of the photon but rather the frequency of atomic oscillators assumed by Max Planck to represent a source of radiation emission.

Electromagnetic, strong, and weak nuclear forces are three of the four fundamental interactions currently known to physicists. The fourth fundamental interaction is the force of gravity. (In section 25 of this book, a fifth fundamental force was identified, and referred to as the force of antigravity.) An electromagnetic force is a long-range force responsible for binding electrons around nuclei, which together form atoms. It is also responsible for binding atoms to form molecules. A strong nuclear force is a very short-range (residual) force responsible for binding protons and neutrons, which form nuclei. A weak nuclear force has the shortest range of all of the fundamental forces. It is responsible for the radioactive decay of subatomic particles (e.g., for beta decay that results in the transformation of protons into neutrons, and vice versa). Like the force of gravity, those three forces play critically important roles in cosmology that relate to the formation of atomic nuclei and chemical elements, and to the evolution of stars affected by the inside nuclear reactions. A detailed understanding of those forces, which are fundamental to the standard model of particle physics, is not necessary for the purpose of this book since neither the formation of atomic nuclei/chemical elements nor the evolution of stars is discussed here.

Entropy is a measure of the number of specific ways in which a thermodynamic system may be arranged. It is commonly understood to be a measure of disorder. Another helpful definition of entropy is that it is a measure of the energy that is not available to do useful work in a closed thermodynamic system.

Epicycles are small planetary trajectory circles invented to support the Ptolemaic (i.e., pre-Kepler) cosmology. For instance, a planet orbiting the sun was assumed in the Ptolemaic cosmology to move along a big circle around the sun, while it also moved along a small circle, the center of which moved along the big circle.

Equivalence principle is the assertion that gravitational and inertial masses are exactly equivalent. In simple terms, that means that a person standing on the earth, who is subject to gravitational acceleration caused by her/his interaction with the mass of the earth, would feel exactly the same as being accelerated to the same acceleration somewhere deep in space, where all mass objects are far away and the gravitational forces are negligible. Originally, the equivalence principle was put forward by Galileo Galilei based on experiments from which he concluded that two different mass objects falling toward the earth under the force of gravity would be subject to identical accelerations.

Flat universe is a term that refers to the intrinsic geometry of the universe in which two parallel lines are neither convergent not divergent—that is, parallel lines will always remain parallel regardless of how far the lines extend. It is a concept introduced in the standard cosmological model. In a flat universe, the sum of a triangle's angles always equals 180 degrees. The geometry of space in a flat universe is flat (Euclidean) in the sense that mass does not curve it to any significant degree, except for local curvatures in the immediate vicinity of massive objects.

Flatness problem refers to the initial mass-energy density of the universe in the big-bang model that would have to have an incredibly precise value right after the big bang so that today's universe could remain essentially flat. The flatness of the universe has been suggested based on astronomic observations. (A more detailed and simply presented explanation of the flatness problem can be found on the following web page: http://en.wikipedia.org/wiki/Flatness_problem.)

Galaxy refers to a system of stars and interstellar gas and dust that is gravitationally bound (e.g., the Milky Way). A typical galaxy may comprise between two hundred and four hundred billion stars.

Galaxy rotation curve is an observed relation between the orbital velocities of the stars in a galaxy and their radial distances to the galaxy center. It is presented as a velocity-versus-distance graph. Such a relation can also be derived from Newton's law of gravitation based on the amount of visible matter in the galaxy. Comparisons of those two relations for the Milky Way and other near-by galaxies indicate that the amount of matter in a galaxy should be roughly ten times higher than the amount of visible matter.

General relativity is Einstein's geometric theory of gravitation. It was developed from the fundamental assumption that curved space-time is a physical reality. The principal ideas of general relativity are often stated as follows: matter tells space-time how to curve, while curved space-time tells matter how to move. However, while curved space-time tells matter how to move, it does tell us why the matter moves—that is, why a mass object is in a free fall. It just tells us that an object in a free fall follows a pathway that is closest to a straight line in the curved space-time. From this perspective, general relativity is a generalization of Newton's law of gravitation that was

Appendix A: Glossary of Terms and Concepts 163

designed to incorporate special-relativity. (In Newton's law of gravitation, an object also falls along a straight line, but it does not affect space.) Because the mathematics of general relativity is much more elaborate than the mathematics of Newton's law of gravitation, general relativity is able to describe gravitational interactions more accurately than Newton's law. Δ

Gradient in physics refers to the "slope" of a function (the slope of the function that shows the magnitude of physical quantity as it varies with another quantity, for example, with distance or time). In simplest terms, "gradient" can be understood to be a difference between two physical quantities divided by the distance between them. For instance, a difference in temperature of 27 degrees between two locations in a body would result in a temperature gradient of $9°/m$ if the two locations were three meters apart. What is of central importance with regard to a gradient in physics is that many physical quantities are proportional to a gradient. For instance, the rate of heat flow between two locations in a body is proportional to the temperature gradient between the two locations. As another example, the rate of the motion of salt ions between two locations in the body of water will be proportional to the salt-concentration gradient between those locations.

Gravitational binding energy is the energy that gravitationally binds matter. Its magnitude is calculated as the amount of energy that would be required to bring a material object into gravitationally unbound state. If the range of gravitational interactions is assumed to be infinite, a fully unbound state would be realized after the matter expands to infinity. The effect of gravitational binding energy is nonnegligible in very massive objects such as stars and planets.

Gravitational lensing refers to gravitational interactions of electromagnetic radiation beams with massive objects such as stars, galaxies, or clusters of galaxies. Those interactions result in the bending of radiation beams when the beams pass close to massive objects. The bending can be observed using optical telescopes. The first and most famous observation of the gravitational bending of a starlight beam passing next to the sun was made in 1919 with the purpose of testing general relativity, from which Einstein predicted the magnitude of the bending.

Gravitational potential energy is the energy that a mass object has because of its position with respect to one or more mass objects. For instance, a stone held in one's hand on a balcony has gravitational potential energy since its kinetic energy will increase as it falls to the ground. The increase in the kinetic energy will be compensated for by the loss of the stone's gravitational potential energy, while its intrinsic energy will remain unchanged.

Gravitational waves are the waves that carry gravitational radiation. Gravitational radiation comprises energy called "gravitational energy." For instance, a star rotating around another star slowly loses its gravitational potential energy and kinetic energy. The sum of those energies represents the gravitational energy of the star. The gravitational energy lost by the star continuously propagates away from the star, throughout space, in the form of a gravitational wave. Typically, such a wave is so feeble that in cannot be detected.

High-z refers to a high value of the redshift (z) of a starlight beam emitted from an astronomic object. That value is calculated from the formula $z = (f_{emitted} - f_{observed})/f_{observed}$, where f is the frequency of the starlight beam. (The value of z is also used to measure astronomic distances (d) according to the formula $d = z c / H_0$, where H_0 is the Hubble's constant.) In the observations of high-z supernova spectra, the redshifts were interpreted as indicative of an accelerating expansion of space. The observed redshifts indicated that the high-z supernovae appeared fainter than they would be if space were decelerating in accordance with the original big-bang model predictions. An ad hoc hypothesis was then introduced that involved the existence of a mysterious "dark energy" spread over the universe that caused its accelerated expansion (rather than the expected deceleration).

Horizon problem refers to the universe being highly homogeneous (i.e., having highly homogeneous distributions of mass-energy and temperatures) in spite of the fact that some parts of the universe could not possibly "communicate" with one another owing to the limit on the speed of information transfer implied by special

Appendix A: Glossary of Terms and Concepts 165

relativity. (A detailed, simply presented explanation of the horizon problem is given on http://en.wikipedia.org/wiki/Horizon_problem.)

Inertial frame of reference is a coordinate system that is in rectilinear (i.e., along a straight line) motion at constant (i.e., nonaccelerated) speed with respect to an observer.

Luminosity (in astronomy) is the rate of the emission of the total electromagnetic energy from a cosmic object.

Olbers's paradox refers to the majority of the dark sky that we observe at night. If the universe were infinite in size, our observable horizon should be filled with starlight, with no dark areas. (Our line of sight would reach a shining star/galaxy in any direction we looked.) That presents an obvious paradox for a spatially-infinite universe only if all the stars/galaxies in such a universe are assumed to shine like the sun.

Paradigm in science refers to the set of practices that define a scientific discipline at a particular time. For instance, Ptolemy's view of the distribution and motions of stars and planets was the astronomic paradigm until Copernicus's work gave rise to a new (scientifically more accurate) paradigm.

Passive galaxy is a galaxy characterized by no, or few, signs of on-going star formation.

Periastron refers to the configuration of a binary-star system in which the two stars are the closest to each other. It also refers to the configuration of a star and an orbiting body in which the body is closest to the star.

Perfect sink refers to an important concept in thermodynamics. Most often, it is used to analyze a thermodynamic system that is subject to a steady-state process. A perfect-sink condition exists at a boundary of such a system at which there is no change, with time, in the value of a thermodynamic variable.

Photon is an elementary particle that is believed to have zero rest mass. As originally proposed by Max Planck in 1900, a photon is a packet, or a quantum of pure energy. It is believed to be always in motion at constant speed. Owing to the so-called wave-particle duality, it is very difficult, if not impossible, to arrive at a satisfactory explanation of what a photon is. (This is the subject of the quantum-mechanics interpretations.) Perhaps the most practical way to think of a photon is that the photon is an electrically neutral, fundamental particle comprising pure energy, while keeping in mind that the propagation of radiation beams, which comprise photons, can be described by the mathematics of a wave equation.

Principle of causality refers to a principle used in physics that states that the cause must precede its effect according to all inertial observers in all inertial frames of reference.

Principle of relativity, which was first stated by Galileo and later used by Einstein in developing special relativity, means that all laws of physics have the same form in all inertial frames of reference.

Quantum refers to the minimum amount of any physical entity, including energy, that can be involved in a physical interaction. Photon as a quantum of light—that is, a quantum of electromagnetic radiation—presents the most common use of the word "quantum."

Quantum entanglement refers to a mysterious phenomenon identified within the realm of quantum physics. A quantum entanglement occurs when two or more objects that interacted with one another at a close distance in the past are still somehow connected (i.e., still somehow interact) even when separated by a very large distance. Most typically, a measurement done on one of the objects is instantaneously correlated with the corresponding measurement done on the other object. The mystery of quantum entanglement could only be removed if the two objects interacted with infinite speed. According to the mainstream physics, however, no interaction can occur faster than the speed of light. Nonetheless, I contend in section 18 that the speed of gravitational interaction should can actually be infinite.

Quantum physics is the branch of physics that describes and attempts to explain matter and the interactions between matter and energy on the scale of subatomic particles. One of the central concepts in quantum physics is the principle of uncertainty, which states that there is a fundamental limit on the accuracy of simultaneous measurements of certain physical quantities such as, for example, momentum and position. Classical physics, on the other hand, deals with matter and energy on the larger scales at which the effect of the principle of uncertainty is negligible.

Appendix A: Glossary of Terms and Concepts 167

Quasars are highly energetic (highly luminous) cosmic objects that are believed to comprise matter and energy contained in small central regions of some massive galaxies.

Radiation frequency (f) is a measure used to describe the energy of radiation (e.g., the energy of sunlight rays). It is defined by the amount of photon energy (E_{ph}) according to the equation $E_{ph} = f/h$, where h is a constant (Planck's constant).

Space-time is a geometric model that combines three dimensions of space and one dimension of time into a four-dimensional continuum. The model is effective in the descriptions of some physical phenomena. The best-known example here is general relativity (a theory of gravitational interactions).

Special relativity is the theory that was developed by Einstein based on the premise of the relativity of lengths and times. Special relativity explained the observed behavior of Maxwell's electric and magnetic fields, without the need to refer to the concept of absolute rest. In particular, Einstein argued that introducing the concept of luminiferous aether, which were to reflect an absolute space, was not necessary. The most celebrated implication of special relativity is the equivalence of mass and energy ($E = mc^2$). Another key implication of special relativity is that the speed of the motion of matter and energy cannot exceed the speed of light.

Standard cosmological model is the most recent version of the big-bang model, which also is called the ΛCDM model. It is based on the fundamental assumption that galaxies recede from one another with speeds proportional to the distances between them. That assumption implies that at some time in the past, all of the matter and energy in the universe (stars, planets, cosmic gas and dust, the energy of electromagnetic radiation, etc.) were confined to an infinitesimally small volume. The standard cosmological model assumes the existence of both dark energy, which causes the accelerated expansion of the universe, and cold dark matter (CDM), which is assumed to exist in order to explain the observed gravitational interactions between stars and galaxies. The symbol "Λ", which represents the cosmological constant originally introduced in the equations of general relativity by Einstein, refers to the dark energy proposition incorporated in the standard model. The other key assumptions underlying the standard model include the existence of space-time as a physical reality; the Doppler-shift interpretation of cosmological redshifts; the relic-radiation interpretation of the cosmic microwave background; the early inflation of the universe; primordial fluctuations (spatial variations of the density of matter-energy in the early universe); the big-bang itself; and the creation of mass-energy, space, time out of nothing at the beginning of time. Neither of those assumptions has ever been verified.

Steady state is the condition of a physical process occurring in a thermodynamic system during which the properties and the behavior of the system at any given location in the system do not change with time.

Supernova (*plural* supernovae or supernovas) refers to a highly energetic, extremely luminous explosion of a massive star. Such an explosion is caused by the gravitational collapse of a massive star and the associated release of the large amount of energy that results from a star's explosion.

Thermal radiation is the electromagnetic radiation emitted from any object that has the surface temperature above absolute zero. The thermal radiation emitted from stars/galaxies has to be one of the most critical features of any cosmological model of the universe. (See also the explanation of "electromagnetic radiation" in this appendix.)

Visible light is radiation that has a frequency between the frequencies of ultraviolet and infrared radiations, in the range of about 400–770 THz (1 THz = 10^{12} Hz).

Void galaxy is a galaxy located in a cosmic void. All void galaxies have been observed to have either spiral or elliptical (i.e., disklike) configurations. The majority of void galaxies have been observed to be "young"—that is, their spectra are "bluer" than those of typical filament galaxies.

White dwarf refers to a very old remnant star that is believed to be in the final evolutionary stage of stars and has a mass comparable to the mass of the sun. The origin of a white dwarf's radiation is the thermal energy stored in the star—that is, energy is no longer generated inside a white dwarf. (Nuclear fusion has ceased.)

APPENDIX B: Index

act of creation; 13, 14, 151
action-at-a-distance; 157, 28-32, 127
aether; 107, 112, 122-123, 125
Berkeley, George; 19, 46
birth of star; xv, 11,13, 126, 139-141
cold-gas accretion; 141, 156
Coles, Peter; 4
cosmic gas and dust; 156, 11, 104, 139, 146
cosmic microwave background (CMB); x, xii, 157, 11, 13, 126, 129, 139, 149, 151-153, 155
cosmic void; 157, 129, 133-137, 139-141, 155, 156
cosmic-void horizon; 136-137, 155
cosmological redshifts; 157, x, xv, 4-7, 8, 125, 129, 144-150, 151, 155, 156
creation of matter; 11, 140, 155
dark energy; 158, ix, 2-3, 8, 104, 138
dark matter; 158, ix, 104, 138-139, 151, 152-153, 156
Davies, Paul; 3, 40, 49
death of star; 11, 13, 126, 129, 139, 141, 143, 155
dehumanization of nature; x, 15-55, 15, 27, 43, 47, 55, 69
Dirac, Paul; 17, 49
distant masses (see also outside masses); 80, 91-92, 95, 136
Doppler shift (see also cosmological redshifts); 158, 4, 5, 7, 144-150, 151
Einstein, Albert; 7, 15, 20, 22, 30-32, 46, 48, 49, 63-64, 65, 70, 95, 100, 103, 109, 110, 111, 112, 113, 122, 125, 136
Einstein's proof (the universe cannot be spatially infinite); 159, 9-10, 56, 136
electrostatic force; viii, 10, 32, 42, 89-91, 93, 98,128
electrostatic-to-gravity-force ratio; viii, 89-95, 97, 98, 128
entropy; 160, 11, 13, 27, 39, 126, 135, 142-143,
equivalence principle; 160, xii, xiv, 23, 55, 98-103, 128, 154
expansion of the universe; vii, xiii, xv, 1-7, 8-9, 12, 13-14, 138, 144

Feynman, Richard; 7, 17, 24, 25, 26, 29, 41, 91
filamentary structure of universe; 129, 133-134, 140-141, 144, 155
finite range of gravity; viii, xi, xiv, 9, 10, 13, 24, 32, 56-60, 63, 69, 72, 76, 78, 83, 95, 100-102, 103, 107, 109, 126, 127, 128, 136, 155
force of antigravity; ix, 96-98, 128, 135
force of gravity; viii, x, xiv, 22, 32, 39, 52, 53, 69, 71, 73, 75, 79-87, 89-91, 93-94, 97-99, 127, 128, 135, 136, 137, 155
force of inertia; 34, 37, 92, 128
fundamental particle; 18, 71-72, 107, 110, 123, 128, 140
fundamental-law hypothesis (FLH); 15, 17-22, 23-25, 26, 27, 29, 33-37, 41-42, 43, 46, 49, 53, 54-55, 56, 64, 71-72, 83, 91, 98, 100, 107-108, 110, 123, 126, 131-132, 127, 139-140
galaxies (clustering of); x, 144
galaxies, void; x, 168, 134, 141, 156
Galileo, Galilei; xiii, 48, 49, 100, 103
general relativity; 161, 1, 7, 9, 16, 20, 22, 23, 30, 41, 48, 55, 56, 57, 64, 71, 92, 95, 100, 111, 154
gravitational potential energy; 163, 32, 57, 62, 68, 71, 111, 128
gravitational redshift; 111, 150
Gravity (gravitational field); xi, xvii, 23, 30, 32, 53, 58-63, 68-79, 81-82, 92, 94, 97, 100, 101, 103-105, 108-110, 111, 112, 113-122, 124-125, 127, 128, 129-135, 136, 137, 140, 141, 154, 155
gravity-conservation law; 23, 24, 76-77, 127
Gravity-free regions; 69, 109-110, 119, 127, 130-134, 140, 141, 155, 156
Greene, Brian; 3
Guth, Alan; ix, 4 10
Hawking, Steven; 3, 10, 16
Hubble, Edwin; 3-6, 124-125, 148, 149, 150
inertia law; 23, 42, 92, 100, 102, 110

infinite space (spatially-infinite universe); vii, 7, 9-10, 11-12, 13, 43, 44, 46, 126, 128, 135, 136, 142
infinite-time hypothesis (infinite time); vii, viii, xiii, 9, 8-10, 11-12, 13, 24, 32, 36, 56, 63, 83, 103, 107, 133, 137, 138, 139, 142, 144, 154
insufficient-mass conclusion; 129-132, 139, 140
intrinsic energy of photon; 108-109, 111, 128, 140
intrinsic mass-energy; viii, 57, 60, 62, 70, 154, 155
invisible stars/galaxies (invisible matter); 12, 13, 126, 137-139, 143, 151, 152, 156
Kaku, Michio; 4
Krauss, Lawrence; 4
Lopez-Corredoira, Martin; 3, 5, 8, 149
Mach, Ernst; 31, 91-92, 95, 128
mass density in gravitational field; 61, 69, 81-83, 84-85, 103-104, 116-117, 127
mass density in the universe; 5, 104, 130, 133, 137-139
mass, amount in universe; 129-132, 137-139, 156
Maxwell, James Clerk; 15, 28-29, 108, 112, 122, 125, 145
Michelson-Morley experiment; 122
Minkowski, Hermann; 7, 30, 48-49, 65-66,
Mitchell, William C.; 8
Newton, Isaac; xiv, 9-10, 12, 29, 30-32, 49, 56, 63, 68-69, 71, 75, 79, 87, 103, 106
Newton's law of gravitation; viii, 10, 21-22, 36, 41, 53, 55, 56, 64, 76, 79, 83, 84, 87, 95, 102, 127, 138
Newton's law of gravitation, modified; x, 79-89, 154
Olbers's paradox; 164, x, 10, 43, 137
outside masses (see also distant masses); 75, 92-95, 124
Penrose, Roger; ix, 3, 49

perfect cosmological principle; vii, 8, 13-14, 47, 126, 130, 153, 154
perpetuum mobile; 142-143
photon (collisions); 77, 110, 140-141,
photon; 165, xv, 6, 7, 12, 70-72, 77, 107-125, 128, 129, 139, 140, 141, 144-150, 151, 152, 153, 155, 156
photon's intrinsic energy; 70, 108-109, 111, 128, 140
photon's kinetic energy; 108-109, 111, 140
physical law; 15, 21, 25, 26, 27, 40-42, 43, 44, 46-47, 48, 50-52, 54, 65, 66, 69, 71, 76, 79, 83, 84, 85, 99, 126, 127, 154
physical reality; xiv, 1, 3, 5, 7-8, 20, 21, 47, 53-54, 65-66, 138, 153
physics and mathematics; 47-54
Planck, Max; 7, 19, 109
principle of relativity; 165, 22, 108
Rowan-Robinson, Michael; 4
simplicity-of-execution principle; 33-37, 56, 79, 80, 83, 84, 91, 98, 100, 107, 127, 154
Slipher, Vesto; 4, 5
space, human; 43, 45, 47, 66
space, physical; 43-47, 127
space-time; 166, xiv, xvii, 7-8, 14, 20, 28, 30, 32, 48, 65-67, 92, 138, 154
standard cosmological (big-bang) model; 166, vii, ix, x, xiii, 1-8, 13, 14, 128, 129, 138, 139, 141, 142, 143, 145, 151, 152, 155
star horizon; 13, 137, 142, 155
steady-state thermal radiation process; 112-117, 119, 128
Stefan-Boltzmann's law; 12, 118
time (human); 1, 8, 43, 45, 47
time (physical); 1, 43-47, 127
vacuum (Newton-Einstein's); 30, 32, 111, 125
vacuum (true); xi, 32, 111, 127
Van Flandern, Tom; 8, 23
Weinberg, Steven; 5, 8, 25, 66
Wilczek, Frank; 3, 49
Zee, Anthony; 4
Zwicky, Fred; 6

APPENDIX C: Bibliography

[1] Alan H. Guth, *Inflationary universe: A possible solution to the horizon and flatness problems*, Phys. Rev. D **23** 347 (1981).
[2] Paul J. Steinhardt, *The Inflation Debate—Is the theory at the heart of modern cosmology deeply flawed?*, Scientific American 38 (April 2011).
[3] Roger Penrose, *Aeons before the Big Bang*, The Second Copernicus Center Lecture, Krakow (2011).
[4] Maciej B. Szymanski, *Gravitation Photons Universe*, Physics Essays **23** 388 (2010).
[5] Maciej B. Szymanski, *The physics of infinite in time universe*, Physics Essays **25** 455 (2012).
[6] Maciej B. Szymanski, *The simple nature of physics*, Physics Essays **25** 590 (2012).
[7] Jean-Claude Pecker, *How to Describe Physical Reality?*, Apeiron **2** 1 (1988).
[8] Andrew Liddle, *An Introduction to Modern Cosmology*, p.xi (John Wiley & Sons, 2008).
[9] Thomas S. Kuhn, *The Structure of Scientific Revolution*, Second Edition (The University of Chicago Press, 1970).
[10] Eds. Martin Lopez-Corredoira & Carlos C. Perelman, *Against the Tide: A Critical Review by Scientists of How Physics and Astronomy Get Done* (Universal Publishers, 1998).
[11] Martin Lopez-Corredoira, *Non-Standard Models and the Sociology of Cosmology*, Studies in History and Philosophy of Modern Physics **46-A** 86 (2014).
[12] Edwin Hubble, *A Relation between Distance and Radial Velocity among Extra-galactic Nebulae*, Proc. N. A. S. **15** 168 (1929).
[13] Roger Penrose, *The Road to Reality*, p.462 (Random House: London, 2004).
[14] Paul Davies, *The Mind of God*, p.51 (Simon & Schuster, 2005).
[15] Stephen Hawking, *A Brief History of Time*, Anniversary Edition, p.9, p.191 (Bantam Books, 1998).
[16] Frank Wilczek, *The Lightness of Being*, p.106 (Basic Books, 2008).
[17] Brian Greene, *The Fabric of Cosmos*, p.229 (Vintage Books, 2005).
[18] Michael Rowan-Robinson, *Exploring the Infrared Universe* (Cambridge University Press, 2013).
[19] Alan Guth, *The Inflationary Universe*, p.20, p.297 (Basic Books, 1997).
[20] Lawrence M. Krauss, *A Universe from Nothing* (Atria Books, 2012).
[21] Michio Kaku, *Physics of the Impossible* (Anchor Books, 2009).
[22] Bernard F. Schutz, *A First Course in General Relativity*, p.338 (Cambridge University Press 2009).
[23] Anthony Zee, *Fearful Symmetry* (Princeton University Press, 2007).
[24] NASA web page: http://map.gsfc.nasa.gov/universe/uni_expansion.html (October 2015).
[25] Peter Coles, *Cosmology—A Very Short Introduction*, p.8 (Oxford University Press 2001).
[26] Vesto M. Slipher, *The Radial Velocity of the Andromeda Nebula*, Lowell Observatory Bulletin **1** 56 (1913).
[27] Edwin Hubble, *The Observational Approach to Cosmology*, p.12 (Oxford University Press 1937).

[28] Reference 27.
[29] Edwin Hubble, *Effects of Red Shifts on the Distribution of Nebulae*, Contributions from the Mount Wilson Observatory, Carnegie Institution of Washington **557** 517 (1953).
[30] Andre K. Assis, Marcos C. D. Neves & Domingos S. L. Soares, *Hubble's Cosmology: From a Finite Expanding Universe to a Static Endless Universe*, Second Crisis in Cosmology Conference, ASP Conference Series **431** 255 (2009).
[31] Martin Lopez-Corredoira, *Observational Cosmology: caveats and open questions in the standard model*, arXiv:astro-ph/0310214v2 (2003).
[32] Steven Weinberg, *The First Three Minutes* (Basic Books, 1993).
[33] Harvard web page: www.cfa.harvard.edu/seuforum/galSpeed/.
[34] Dainis Dravins, Lennart Lindegren & Soren Madsen, *Astrometric radial velocities*, Astron. Astrophys. **348** 1040 (1999).
[35] Fred Zwicky, *On the Red Shift of Spectral Lines Through Interstellar Space*, Proceedings of the National Academy of Sciences **15** 773 (October 1929).
[36] Paul A. Violette, *Is the Universe Really Expanding*, Astrophysical Journal **301** 544 (1986).
[37] J. C. Pecker, A. P. Roberts & J. P. Vigier, *Non-velocity red shifts and photon-photon interactions*, Nature **237** 227 (1972).
[38] J. C. Pecker & J. P. Vigier, *A possible tired-light mechanism*, Apeiron **2** 19 (1988).
[39] Michael Harney, *The Cosmological-Redshift Explained by the Intersection of Hubble Spheres*, Apeiron **2** 288 (2006).
[40] Jose Francisco Garcia Julia, *Simple Considerations on the Cosmological Redshift*, Apeiron **15** 325 (2008).
[41] Amitabha Gosh, *Velocity-Dependent Inertial Induction*, Apeiron **9–10** 95 (1991).
[42] Richard P. Feynman, *QED—the strange theory of light and matter* (Princeton University Press, 1985).
[43] Max Planck, October 1900 presentation to Deutsche Physikalische Gesellschaft.
[44] Albert Einstein, *Uber einen die Erzeugung und Verwan dlung des Lichtes betreffenden heuristischen Gesichtspunkt* (*Concerning an Heuristic Point of View Toward the Emission and Transformation of Light*), Ann. Phys. **17** 132 (195).
[45] H. Minkowski, Address delivered at the Eightieth Assembly of German Natural Scientists and Physicians (1908) English translation in *The Principle of Relativity*, p. 73 (Dover Publications, 1952).
[46] Steven Weinberg, *Gravitation and Cosmology: Principles and Applications of the General Theory of Relativity*, p.vii (John Wiley & Sons, 1972).
[47] Tom van Flandern, *The Top 30 Problems with the Big Bang*, Meta Research Bulletin **11** 6 (2002).
[48] William C. Mitchell, *The cult of the big bang: Was there a bang?* (Cosmic Sense Books, 1995).
[49] Reference 31.
[50] Michael J. Disney, *Doubts About Big Bang Cosmology*, in Antonio Alfonso-Faus (Ed.) 'Aspects of Today's Cosmology' (2011).
[51] Yurij Baryshev, *Paradoxes of cosmological physics in the beginning of the 21-st century*, arXiv:1501.01919v1 (2015).

Appendix C: Bibliography

[52] Andre K. T. Assis, *A Steady-State Cosmology* in 'Progress in New Cosmologies: Beyond the Big Bang' edited by Halton C. Arp, C. Roy Key & Konrad Rudnicki (Plenum Press 1993).

[53] Paul J. Steinhardt & Neil Turok, *Cosmic Evolution in a Cyclic Universe*, Phys. Rev. D **65** 126003 (2002).

[54] Albert Einstein, *Relativity: The Special and General Theory*, p.97 (Penguin Books, 2006; originally published by Henry Holt: New York, 1920).

[55] Reference 19.

[56] Reference 15, p.6.

[57] Bradley Dowden, *Time* (Internet Encyclopedia of Philosophy, 2010).

[58] Sean M. Carroll, *From Eternity to Here: The Quest for the Ultimate Theory of Time* (Penguin Group, 2010).

[59] Arthur Eddington, *The Philosophy of Physical Science* (Cambridge University Press, 1939).

[60] Craig Callender & Carl Hoefer, *Philosophy of Space-Time Physics* in *The Blackwell Guide to the Philosophy of Science*, ch. 9 (Blackwell Publishers Ltd. 173, 2002).

[61] Holger Lyre, *Time in Philosophy of Physics: The Central Issues*, Physics and Philosophy (Technische Universität Dortmund, Issn: 1863-7388 Id: 012, 2008).

[62] Craig Callender, *Is Time an Illusion?*, Scientific American 59 (May 2010).

[63] Mark Colyvan, *The Miracle of Applied Mathematics*, Synthese **127**, 265 (2001).

[64] Reference 15, p.7.

[65] Stephen Hawking & Leonard Mlodinow, *The Grand Design* (Bantam Books: New York, 2010).

[66] Paul Adrien Maurice Dirac, *The Relation Between Mathematics and Physics*, Proceedings of the Royal Society (Edinburgh) **59** 112–129 (1940).

[67] Interview with Max Planck by J. W. N. Sullivan, published in The Observer (25 January 1931).

[68] Eugene P. Wigner, *Remarks on the Mind-Body Question*, published in *The Scientist Speculates*, ed. I. J. Good, 284 (Heinemann: London, 1961).

[69] Greard 't Hooft, *The Obstinate Reductionist's Point of View on the Laws of Physics*, Lecture given at the 2001 Technology Forum, Alpbach, Austria (2001).

[70] Tom Van Flandern, *The Speed of Gravity—What the experiments say*, Physics Letters A **1-3** 1 (1998).

[71] T. A. Wagner, S. Schlamminger, J. H. Gundlach & E. G. Adelberger, *Torsion-balance tests of the weak equivalence principle*, Class. Quantum Grav. 29 184002 (2012).

[72] Steven Weinberg, *Towards the Final Laws of Physics* (The 1986 Dirac Memorial Lectures, Cambridge University Press, 1987).

[73] Richard P. Feynman, *The Character of Physical Law*, p.57 (Random House: NY, 1965).

[74] Joseph Melia, Mind **109** 435 (2000).

[75] Julian B. Barbour, *The Emergence of Time and Its Arrow from Timelessness*, published in *Physical Origins of Time Asymmetry*, eds. J. J. Halliwell, J. Perez-Mercader & W. H. Zurek (Cambridge University Press, 1994).

[76] Richard P. Feynman, *The Feynman Lectures of Physics*, Vol. (Addison-Wesley Publishing Co: Reading, 1977).

[77] Eds. Andrew E. Chubykalo, Viv Pope & Roman Smirnov-Rueda, *Instantaneous Action at a Distance in Modern Physics: Pro and Contra* (Nova Science Publishers: Commack, 1999).

[78] Jayant V. Narlikar, *Action at a Distance in Electrodynamics and Inertia*, published in *Instantaneous Action at a Distance in Modern Physics: Pro and Contra*, eds. Andrew E. Chubykalo, Viv Pope & Roman Smirnov-Rueda (Nova Science Publishers: Commack, 1999).

[79] Andrew E. Chubykalo & Roman Smirnov-Rueda, *Action at a distance as a full-value solution of Maxwell's equations*, arXiv:hep-th/9510052v2.

[80] Archibald Wheeler & Richard P. Feynman, *Interaction with the Absorber as the Mechanism of Radiation*, Rev. Mod. Phys. **17** 157 (1945).

[81] Fred Hoyle & Jayant Narlikar, *Cosmology and action-at-a-distance electrodynamics*, Rev. Mod. Phys. **67** 113 (1995).

[82] Isaac Newton's letter to Richard Bentley (Feb. 25, 1692).

[83] Isaac Newton, *The Mathematical Principles Of Natural Philosophy (Principia)*, Book III, General Scholium, p.506 (translated by Andrew Motte, published by Daniel Adee, New York, 1846).

[84] John Henry, *Gravity and De gravitatione: the development of Newton's ideas on action at a distance*, Studies in History and Philosophy of Science **42** 11 (2011).

[85] Albert Einstein, *Die Grundlage der allgemeinen Relativitätstheorie* (On the Foundation of the General Theory of Relativity), Annalen der Physik **7** 769 (1916) (English translation in *The Principle of Relativity*, Dover Publications, Inc., 1952).

[86] Reference 54, p.60.

[87] Albert Einstein's letter to Max Born (March 3, 1947).

[88] Albert Einstein's letter to Michele Besso (August 10, 1954).

[89] Ernst Mach, *History and Root of the Principle of the Conservation of Energy* (The Open Court Publishing Co.: Chicago, 1911) (English translation by Philip E. B. Jourdain, Microfilm-Xerox by University of Microfilms Inc.: Ann Arbor, 1959).

[90] G. M. Wang, E. M. Sevick, Emil Mittag, Debra J. Searles & Denis J. Evans, *Experimental Demonstration of Violations of the Second Law of Thermodynamics for Small Systems and Short Time Scales*, Phys. Rev. Lett. **89** 5 050601 (2002).

[91] A. Brian Pippard, *The Elements of Classical Thermodynamics* (Cambridge University Press, 1957).

[92] Reference 14, p.81.

[93] Henri Poincaré & George B. Halsted, *The Value of Science* (New York The Science Press, 1907).

[94] Reference 73, p.33.

[95] Norman Swartz, *Laws of the nature* (Stanford Encyclopedia of Philosophy, 2009).

[96] Lee Smolin, *The Trouble with Physics* (Mariner Books, 2007).

[97] Aristotle, *Physics*, Book 4, Chapter 11 (Translated by R. P. Hardie & R. K. Gaye, 350 BC).

[98] Jeff McDonough, *Leibniz's Philosophy of Physics* (Stanford Encyclopedia of Philosophy, 2007).

[99] George Berkeley, *A Treatise Concerning Principles of Human Knowledge*, §CXII (Jacob Thonson: London, 1734).

Appendix C: Bibliography

[100] Albert Einstein, *Zur Elektrodynamik bewegter Körper* (On the Electrodynamics of Moving Bodies), Annalen der Physik **322** 891 (1905).

[101] Eugene Wigner, *The Unreasonable Effectiveness of Mathematics in the Natural Sciences*, Comm. Pure. Appl. Math. **13** 1 (1960).

[102] Patricia Fara, *Science: A Four Thousand Year History*, p.24-25 (Oxford University Press, 2009).

[103] Web page: http://www-groups.dcs.st-and.ac.uk/~history/Biographies/Minkowski.html.

[104] Lewis Pyenson, *Hermann Minkowski and Einstein's special theory of relativity*, Arch. History Exact Sci. **17** 71 (1977).

[105] Paul Dirac, *The Evolution of the Physicist's Picture of Nature*, Scientific American **5** (1963).

[106] Reference 101.

[107] Reference 13, p.9.

[108] Frank Wilczek, *Reasonably effective: I. Deconstructing a miracle*, Physics Today **58** 8 (2006).

[109] Paul Davies, *The Mind of God*, p.140 (Simon & Schuster, 2005).

[110] Albert Einstein, *Geometry and Experience*, 21 January 1921 address at the Prussian Academy of Sciences (Methuen & Co. Ltd.: London, 1922).

[111] Mark Steiner, *The Applicability of Mathematics as a Philosophical Problem* (Harvard University Press, 1998).

[112] Willard van Quine, *Two Dogmas of Empiricism*, published in 1951; reprinted in *From a Logical Point of View* (Harvard University Press, 1980).

[113] Hilary Putnam, *Philosophy of Logic* (Harper and Row: New York, 1971).

[114] Mark Colyvan, *Indispensability Arguments in the Philosophy of Mathematics* (Stanford Encyclopedia of Philosophy, 2011).

[115] Alan Baker, *Are there Genuine Mathematical Explanations of Physical Phenomena?*, Mind **114** 223 (2005).

[116] John Burgess, *Why I am not a nominalist.*, Notre Dame Journal of Formal Logic **24** 93 (1983).

[117] Bob Hale & Crispin Wright, *Nominalism and the Contingency of Abstract Objects*, Journal of Philosophy **89** 111 (1992).

[118] Mark Colyvan, *The Indispensability of Mathematics* (Oxford University Press, 2001).

[119] **Mark Steiner, *Mathematics, Explanation and Scientific Knowledge*, Nous 12 17 (1978).**

[120] Hartry Field, *Science Without Numbers: A Defense of Nominalism* (Princeton University Press, 1980).

[121] Elliott Sober, *Mathematics and Indispensability*, The Philosophical Review **102** 35 (1993).

[122] Peter Woit, *The Problem with Physics*, Cosmos **16** 1 (2007).

[123] Reference 96.

[124] Isaac Newton, *Philosophiae Naturalis Principia Mathematica*, Book III Propositions I-XIII (English translation by Andrew Motte, Daniel Adee: New York, 1846).

[125] Allen D Allen, *Finite Gravity: From the Big Bang to Dark Matter*, Int. Journal of Astronomy and Astrophysics **3** 180 (2013).

[126] Peter G. O. Freund, Amar Maheshwari & Edmond Shonberg, *Finite-Range Gravitation*, The Astrophysical Journal **157** 857 (1969).

[127] S.V. Babak & L.P. Grishchuk, *Finite-Range Gravity and Its Role in Gravitational Waves, Black Holes and Cosmology*, International Journal of Modern Physics D **12** 1905 (2003).

[128] Steven Carlip, *Kinetic Energy and the Equivalence Principle*, arXiv:gr-qc/9909014v1 (1999).

[129] Robert Hooke, *An Attempt to Prove the Motion of the Earth by Observations*, Cutler Lecture (The Royal Society, 1674).

[130] Henri Poincaré, *Mathematics and Science: Last Essays* (Dover Publications: New York, 1963), originally published in 1913 by Ernest Flammarion in French.

[131] N. David Mermin, *What's bad about this habit*, Physics Today **62** 8 (May 2009).

[132] Eds. Steven Weinberg in P. C. W. Davies & J. Brown, *Superstrings: A Theory of Everything*, pp.21–24 (Cambridge University Press, 1988).

[133] E. B. Fomalont & S. M. Kopeikin, *The Measurement of Light Deflection from Jupiter*, Astrophys. J. **598** (2003).

[134] Clifford M. Will, *Propagation Speed of Gravity and the Relativistic Time Delay*, Astrophys. J. **590** 683 (2003).

[135] Steven Carlip, *Model-Dependence of Shapiro Time Delay and the "Speed of Gravity/Speed of Light" Controversy*, Class. Quant. Grav. **21** (2004).

[136] Stuart Samuel, *On the Speed of Gravity and the Jupiter/Quasar Measurement*, Int. Journal of Modern Physics D **09** (2004).

[137] J. M. Weisberg & J. H. Taylor, *Relativistic Binary Pulsar B1913+16: Thirty Years of Observations and Analysis*, arXiv:astr-ph0407149v1.

[138] B. P. Abbott et al., *Observation of Gravitational Waves from a Binary Blackhole Merger*, Phys. Rev. Lett. **116** 061102 (2016).

[139] Bernard F. Schutz, *Gravitational waves on the back of an envelope*, Am. J. Phys. **52** 412 (1984).

[140] Albert Einstein, *Über den Einfluss der Schwerkraft auf die Ausbreitung des Lichtes* (On the Influence of Gravitation on the Propagation of Light), Annalen der Physik **35** 898 (1911) (English translation in *The Principle of Relativity*, Dover Publications, 1952).

[141] Albert Einstein, *Ist die Trägheit eines Körpers von seinem Energiegehalt abhängig?* (Does the Inertia of a Body Depend upon its Energy-Content?), Annalen der Physik **18** 639 (1905) (English translation in *The Principle of Relativity*, Dover Publications, 1952).

[142] Isaac Newton, *Opticks: or, a Treatise of the Reflections, Refractions, Inflections and Colours of light*, The Third Books Of Opticks – Query 1 (Printed for Sam. Smith & Benj. Walford, at the Prince's Arms in St. Paul's churchyard, 1704).

[143] John A. Wheeler, *Geons, Black Holes, and Quantum Foam: A Life in Physics* (W.W. Norton & Co.: New York, 1998).

[144] Reference 73, p.31.

[145] Ernst Mach, *The Science of Mechanics* (The Open Court Publishing Co.: Chicago, 1893, reprint, 1919).

[146] D. W. Sciama *On the Origin of Inertia*, Monthly Notices of the Royal Astronomical Society **113** 34 (1953).

[147] C. Brans & R. H. Dicke, *Mach's Principle and a Relativistic Theory of Gravitation*, Physical Review **124** 925 (1961).

[148] Albert Einstein, *The Meaning of Relativity* (Princeton University Press, 1923).

[149] D.W. Sciama, *The Unity of the Universe* (Doubleday & Company Inc., 1959).

[150] Reference 71.
[151] Reference 4.
[152] Asaph Hall, *A Suggestion in the Theory of Mercury*, Astronomical Journal **319** 49 (1894).
[153] NASA web page: http://map.gsfc.nasa.gov/universe/uni_matter.html (October 2015).
[154] Mike J. Disney, *The Case Against Cosmology*, General Relativity and Gravitation, **32** 1125 (2000).
[155] Reference 83, p.506.
[156] Reference 141.
[157] Albert Einstein, *The Meaning of Relativity* (Princeton University Press, 1923).
[158] Reference 141.
[159] Albert Michelson & Edward Morley, *On the Relative Motion of the Earth and the Luminiferous Ether*, American Journal of Science **34** 333 (1887).
[160] Michel Janssen & John Stachel, *The Optics and Electrodynamics of Moving Bodies*, Max Planck Institute for the History of Science, Preprint **265** (2004).
[161] Reference 100.
[162] Reference 27, p.26.
[163] M. Joeveer, J. Einasto & E. Tago, *Spatial distribution of galaxies and of clusters of galaxies in the southern galactic hemisphere*, MNRAS **185** 357 (1978).
[164] S. A. Gregory & L. A. Thompson, *The Coma/A1367 supercluster and its environs*, Astrophysical Journal **222** 784 (1978).
[165] R. P. Kirshner, A. Oemler Jr., P. L. Schechter & S. A. Schectman, Astrophysical Journal **248** L57 (1981).
[166] J. Huchra, M. Davis, D. Latham & J. Tonry, Astrophysical Journal, Supplement **52** 89 (1983).
[167] J. R. Bond, L. Kofman & D. Pogosyan, *How Filaments Are Woven into the Cosmic Web*, Nature **380** 603 (1966).
[168] Rien van de Weygaert & Erwin Platen, *Cosmic Voids: Structure, Dynamics and Galaxies*, Int. Journal of Modern Physics: Conference Series **1** 41 (2011).
[169] E. G. Patrick Bos, Rien van de Weygaert, Klaus Dolag & Valeria Pettorino, *The darkness that shaped the void: dark energy and cosmic voids*, MNRAS **426** 440 (2012).
[170] Jounghun Lee & Daeseong Park, *Rotation of Cosmic Voids and Void Spin Statistics*, Astrophysical Journal **652** 1 (2006).
[171] Jeremy L. Tinker & Charlie Conroy, *The Void Phenomenon Explained*, Astrophysical Journal **691** 633 (2009).
[172] Caroline Foster & Nelson A. Lorne, *The Size, Shape, and Orientation of Cosmological Voids in the Sloan Digital Sky Survey*, Astrophysical Journal **699** 1252 (2009).
[173] Danny C. Pan, Michael S. Vogeley, Fiona Hoyle, Yun-Young Choi & Changborn Park, *Cosmic Voids in Sloan Digital Sky Survey Data Release 7*, arXiv:1103.4156v2 (2011).
[174] Ulrich Lindner, Jaan Einasto, Maret Einasto, Wolfram Freudling, Klaus Fricke & Erik Tago, *The Structure of Supervoids—I: Void Hierarchy in the Northern Local Supervoid*, arXiv:astro-ph/9503044 (1995).
[175] Jeffrey Bennet, Megan Donahue, Nicholas Schneider & Mark Voit, *The Essential Cosmic Perspective* (Pearson Addison-Wesley, 2007).

[176] John Bahcall, Tsvi Piran & Steven Weinberg, *Dark Matter in the Universe* (World Scientific, 2004).

[177] Neta A. Bahcall & Andrea Kulier, *Tracing mass and light in the Universe: where is the dark matter?*, arXiv:1310.0022v2 [astro-ph.CO] (2014).

[178] R. van de Weygaert, K. Kreckel, E. Platen, B. Beygu, J. H. van Gorkom, J. M. van der Hulst, M. A. Arag'on-Calvo, P. J. E. Peebles, T. Jarrett, G. Rhee, K. Kova˘c, C.-W. Yip, *The Void Galaxy Survey*, arXiv:1101.4187v1 (2011).

[179] Randall R. Rojas, Michael S. Vogeley, Fiona Hoyle & Jon Brinkmann, *Photometric Properties of Void Galaxies in the Sloan Digital Sky Survey*, arXiv:astro-ph/0307274v2 (2004).

[180] David M. Goldberg, Timothy D. Jones, Fiona Hoyle, Randall R. Rojas, Michael S. Vogeley & Michael R. Blanton, *The Mass Function of Void Galaxies in the SDSS Data Release 2*, Astrophysical Journal **621** 643 (2005).

[181] K. Kreckel, E. Platen, M.A. Aragon-Calve, J.H. van Gorkom, R. van de Weygaert, J. M. van der Hult & B. Beygu, *The Void Galaxy Survey: Optical Properties and HI Morphology and Kinematics*, arXiv:1204.5185v1 (2012).

[182] E. Ricciardelli, A. Cava, J. Varela & V. Quilis, *The star formation activity in cosmic voids*, MNRAS **445** 4045 (2014).

[183] Pascal A. M. de Theije, Peter Kartgert & Eelco van Kampen, *The Shapes of Galaxy Clusters*, Mon. Not. R. Astron. Soc. **273** 30 (1995).

[184] Jayant V. Narlikar, *Noncosmological redshifts*, in Quasars, IAU Symp., Reidel 463 (1986).

[185] Halton Arp, Christopher Fulton & Daniele Carosati, *Intrinsic Redshifts in Quasars and Galaxies*.

[186] Halton Arp, *Evolution of Quasars into Galaxies and Its Implications for the Birth and Evolution of Matter*, Apeiron **5** 135 (1998).

[187] Jean-Claude Pecker, *Local abnormal redshifts*, in 'Current Issues in Cosmology' edited by Jean-Claude Pecker and Jayant V. Narlikar (Cambridge University Press 2006).

[188] Reference 42.

[189] S. Perlmutter, G. Aldering, G. Goldhaber, R.A. Knop, P. Nugent, P. G. Castro, S. Deustua, S. Fabbro, A. Goobar, D. E. Groom, I. M. Hook, A. G. Kim, M. Y. Kim, J. C. Lee, N. J. Nunes, R. Pain, C. R. Pennypacker, R. Quimby, *Measurements of Ω and Λ from 42 High-Redshift Supernovae*, The Astrophysical Journal **517** 565 (1999).

[190] Halton Arp, *Seeing Red: Redshifts, Cosmology and Academic Science* (Apeiron, 1998).

[191] Martin Lopez-Corredoira & Carlos M. Gutiérrez, *The field surrounding NGC 7603: cosmological or non-cosmological redshifts?*, Astronomy & Astrophysics **421** 407 (2004).

[192] Reference 185.

[193] A. K. T. Assis & M. C. D. Neves, *History of the 2.7 K Temperature Prior to Penzias and Wilson*, Apeiron **2** (3) (1995).

[194] Geoffrey Burbidge, *The Stage of Cosmology*, in 'Current Issues in Cosmology' edited by Jean-Claude Pecker and Jayant V. Narlikar (Cambridge University Press 2006).